HUMAN GENOME RESEARCH AND THE CHALLENGE OF CONTINGENT FUTURE PERSONS

HUMAN GENOME RESEARCH AND THE CHALLENGE OF CONTINGENT FUTURE PERSONS

Toward an Impersonal Theocentric Approach to Value

JAN CHRISTIAN HELLER

CREIGHTON UNIVERSITY PRESS
Omaha, Nebraska
Association of Jesuit University Presses

Library of Congress Catalog Card Number: 95-083920
ISBN: 1-881871-19-3 cloth
1-881871-20-7 paper

EDITORIAL
Creighton University Press
2500 California Plaza
Omaha, Nebraska 68178

MARKETING & DISTRIBUTION
Fordham University Press
University Box L
Bronx, New York 10458

Printed in the United States of America

To
Linda and Abigail

CONTENTS

ACKNOWLEDGMENTS

Robert L. DeHaan, PhD, W. P. Timmie Professor of Anatomy and Cell Biology and first Director of the Center for Ethics in Public Policy and the Professions at Emory University, provided a pre-doctoral fellowship through the Center from 1990 through 1992 which enabled me to investigate the Human Genome Project's ethical, legal, and social implications. He later served on my dissertation committee, and has never failed to encourage and challenge my thinking. Ms. Kathy Kinlaw, Associate Director of the Center, introduced me to Professor DeHaan and has proved to be a true friend. Also while serving at Emory's Ethics Center, I met Peter H. Aranson, PhD, Chair of the Department of Economics, who patiently introduced me to rational choice theory in the context of science and technology policy, and Nick Fotion, PhD, Professor of Philosophy, who introduced me to the work of Derek Parfit and became a friend and colleague in the process.

I also wish to thank the other members of my dissertation committee. E. Clinton Gardner, PhD, Professor Emeritus of Christian Ethics, graciously served as adviser when he could have been enjoying other, well-earned pursuits of retirement. He gave many hours to reading and editing drafts of the dissertation, and consistently encouraged me to find my own voice. Jon P. Gunnemann, PhD, Director of the Graduate Division of Religion and Professor of Social Ethics, read several early drafts and, as always, asked the right questions. Steven M. Tipton, PhD, Professor of Sociology of Religion, and John Snarey, PhD, Professor of Human Development and Ethics, helped to make the defense as humane as possible.

To James M. Gustafson, PhD, Henry R. Luce Professor of Humanities and Comparative Studies, I owe more than I can tell. I first read his two-volume *Ethics from a Theocentric Perspective* in the quiet solitude of Holy Cross Monastery, overlooking the Hudson River. Later, at Emory, we discussed his work and my own over numerous bag lunches. He invariably listened with interest and suggested helpful ways to proceed. I will always be grateful for the opportunity to have studied under him.

Sr. Jane Gerety, RSM, PhD, Senior Vice President for Sponsorship at Saint Joseph's Health System in Atlanta, also deserves a special note of thanks. She has been constant in her support and encouragement. And the quiet efficiency of Ms. Joan Frost, our Executive Secretary, makes everything easier.

I also want to thank Ruth Purtilo, PhD, Director of the Center for Health Policy and Ethics at Creighton University, who introduced the manuscript to Creighton University Press, and Brent Spencer, PhD, Editor at Creighton University Press, who brought it to press.

Lastly, I must thank my wife, Linda, without whose support and encouragement I could not have endured, and acknowledge our daughter, Abigail, who was born during the second year of my graduate studies and provided the inspiration for a book on future generations.

My daddy run off from home when he was fifteen. Otherwise I'd of been born in Alabama.

You wouldnt of been born at all.

What makes you say that?

Cause your mama's from San Angelo and he would never of met her.

He'd of met somebody.

So would she.

So?

So you wouldnt of been born.

I dont see why you say that. I'd of been born somewheres.

How?

Well why not?

If your mama had a baby with her other husband and your daddy had one with his other wife which one would you be?

I wouldnt be neither of em.

That's right...

You're makin my goddamn head hurt.

I know it. I'm makin my own.

Cormac McCarthy, *All the Pretty Horses*

1

A Two-Pronged Problem

> If a full-scale [human genome] project were to be undertaken, its long-range impacts and consequences for future generations would need to be considered as part of an ethical analysis.[1]
>
> Mark Lappé

> The trouble with future people, we might say, is not that they do not exist yet, it is that they might not exist at all. Further, what and how many future people will exist depends upon the decisions and actions of present people.[2]
>
> Gregory Kavka

The US Human Genome Project (HGP) holds the potential to alter dramatically the way scientific medicine is both conceptualized and practiced. The largest project of its kind and the first "big science" project in the history of biological research, HGP is a fifteen year, three billion dollar, basic biological research effort sponsored by the US federal government. It began in 1990 and its goal is to produce a set of "maps" of the entire human genome (that is, of all the genetic material in a nucleated human cell), and of several other model organisms. Conceptually, researchers believe that these maps will lead to an entirely new approach to human disease, what Victor McKusick calls a "new paradigm" for scientific medicine.[3] It is a reductive paradigm, one that is aimed at understanding and addressing on the molecular level the root biological causes of disease. In terms of medical practice, HGP will spin off large numbers of applied research and development efforts that are, in turn, expected to lead to hundreds, and perhaps thousands, of new diagnostic and therapeutic clinical applications. As these clinical applications enter routine or mainstream medicine, they will be used for the prevention, detection, and treatment of thousands of known genetic diseases, and of many other diseases for which individuals are genetically predisposed.

These optimistic expectations notwithstanding, HGP's funding and implementation raised concerns in several quarters. As we will see in more detail below, much of this concern initially came from the scientific community itself, and was focused on the "big science" or "dedicated" shape of the project. Other concerns originated with informed segments of the public, and were focused less on the project as such and more on the adverse effects that HGP's research might hold for society generally. To address these concerns, which had quickly become the subject of several congressional hearings and of numerous articles in the popular and scientific press, the project's organizers proposed a subsidiary research program. This program is called the Ethical, Legal, and Social Implications (ELSI) Program and it is located administratively under HGP. ELSI is sponsoring research by social scientists, legal scholars, philosophers, and even theologians that is intended to anticipate HGP's future implications for society and to forestall or minimize its foreseen adverse effects with policy suggestions and educational efforts. My own interest in HGP took shape as I began to investigate ELSI's first list of ethical, legal, and social implications expected of HGP, following its publication in 1990.[4] This list, as we will see below, is comprehensive. At the time, however, I was interested less in the particular problems it identifies and more in what it seems to assume.

Most of the items on ELSI's first list are concerned with the implications of HGP's expected clinical applications, as opposed to the basic research project itself. For example, as HGP's diagnostic and therapeutic applications enter routine medical practice, there is great concern that the sheer volume of information they produce will make it nearly impossible to protect patient confidentiality and privacy relative to interested third parties, such as employers and insurance companies. Also, even though HGP is intended to help close a long-standing "therapeutic gap" in clinical genetics, the research and development "sequelae" predicted for its applications suggest that the project may significantly aggravate this gap, at least in the near-term. On a philosophical level, there is also concern that the success of HGP's reductive paradigm will be over-generalized by poorly informed medical professionals and patients, adversely affecting views of human determinism and responsibility.

These concerns suggest that ELSI's Working Group assumes that HGP's clinical applications will be the principal source of the problems arising with or on the basis of HGP. Initially, I suspected that this assumption might be politically motivated by the need to deflect public attention away from the project as such; however, I have concluded that even if it is politically motivated, ELSI is probably focused correctly in any case. It is as these applications are used or misused that the public will be most directly affected by HGP's research. Nevertheless, I will argue that a focus on the project's expected applications should not lead us to think that the project as such can be ignored. The project's basic science establishes the conditions that permit its clinical applications to be developed, and below I will outline how observers believe that HGP's so-called "big science" shape largely determines the problematic research and development "sequelae" expected of its applications.

This said, ELSI's first list of expected ethical, legal, and social implications also reflects an implicit assumption, one that led to the development of my own concerns relative to HGP. I became interested in the fact that the majority of HGP's clinical applications will be developed in the *future*, and many perhaps in the rather distant future. This suggests that ELSI also assumes (perhaps uncritically, since it is not mentioned on the list) that HGP's intended *beneficiaries* will live primarily in the future. It was in investigating the significance of this implication that the book's basic problematic took shape.

HGP AND FUTURE GENERATIONS

The concern I attempt to address in the book is two-pronged. It can be stated in a straightforward way, though addressing it is a rather more complicated affair. I am interested, first, in investigating *how* HGP is likely to affect future generations, positively and negatively. This is largely a descriptive and predictive problem.[5] The second problem is largely prescriptive. I am interested in asking what implications the project's effects on future generations have for *evaluating* HGP and far-reaching research efforts like it, particularly from a theological perspective. Let me hasten to add that I am not attempting to evaluate the overall worthiness of HGP, which would be a vastly more difficult project; rather, I am asking how its effects on future generations ought to figure into such an overall evaluation. This effort is complicated enough.

As we will see, it is an effort that is beset with uncertainty, both empirical and ethical.

With respect to the first prong of the problem—that is, its descriptive and predictive aspects—I will claim that HGP's applied products are more likely to benefit persons living in the *future* than persons living now. This may or may not be seen as a controversial claim, depending on one's perspective. It is based on predictions made by a number of prominent scientists to the effect that the new knowledge produced by HGP's fifteen years of research will be of such a quality and quantity that it will keep researchers busy for perhaps hundreds of years interpreting it and developing applications on the basis of it. This prediction could, of course, be proven wrong in at least two ways. Funding could be interrupted, and anticipated applications might never be developed for this reason; or the project could be more successful than even its most optimistic early supporters anticipated, and its applications developed much more quickly than expected. Most informed observers seem to believe, however, that something between these two extremes is more likely to happen. They believe that HGP will spin off a large number of smaller clinical application projects, and that the scope and the complexity of the research will combine with funding limitations to make this a multi-generational effort.

Now, while I am persuaded by these observers, this predictive uncertainty nevertheless illustrates a major concern associated with evaluating any far-reaching effort like HGP. Such an evaluation is complicated by the question of whether the empirical uncertainties inherent in it can be sufficiently overcome in principle and in practice to give it validity. In the case of HGP, I believe that predictions sufficient for my purposes are possible, and below I suggest ways that our evaluative methods can be qualified so as to lessen the empirical burdens of prediction. Thus, on the basis of this predictive effort, I will outline a probable three-phase *inter-generational* allocation pattern that is anticipated to result from HGP's spin-off research and development efforts. With respect to evaluating HGP's effects on future generations, observers believe that phase two generations will experience considerably higher costs and risks of harm. I suggest that this allocation pattern raises an interesting question concerning what is most frequently discussed as "justice between generations" or as "inter-generational justice."

However, before this question can be addressed, we must consider a second and more perplexing way HGP will affect future generations. Interestingly, this second way holds even if the above predictions concerning HGP's future allocation pattern are proved wrong. It concerns a puzzling relation that at least one of HGP's applications, the preimplantation diagnosis of genetic disease, already holds toward some future people. I also consider a second application, germ-line gene therapy, which could hold this puzzling relation as well if it is developed for use in humans. Due to applications such as these (and perhaps others not considered here, such as human cloning and embryo reduction based on genetic diagnosis), some future persons will exist only *because* of HGP's research. That is, the very existence of some future persons and, by extension, their numbers and identities, will depend on or, as I typically say here, will be "contingent" on the development and use of applications made possible by HGP's research.[6] By claiming that the existence, numbers, and identities of some future persons are contingent on some of HGP's applications, I intend to imply that these persons are contingent in the sense that their lives will depend in ways never before possible technically on specific human choices, but not in the more general sense that their existence, numbers, and identities depend on chance events. All persons are contingent in this latter sense. Now, what is *ethically* interesting about these future persons is that they cannot meaningfully be said to be harmed or benefited by the applications on which their existence depends. This claim constitutes what I call the "problem of contingent future persons," and it poses particularly vexing challenges for evaluating such applications and the research policy that makes them possible. We will see that addressing the problem takes us into some very abstract, even esoteric, kinds of considerations in the second half of the book. Thus, before we move into these considerations, I wish to illustrate how the problem of contingent future persons arises with HGP's applications, using two hypothetical cases.

Two Illustrative Cases

The first case illustrates the preimplantation diagnosis of genetic disease, an application that utilizes a set of new diagnostic techniques emerging both from HGP's research and from other, more traditional research in molecular genetics and human reproduction. The second case uses what is called germ-line gene therapy to illustrate a related choice that the same clients might have available to them in the near future, if current federal regulations against germ-line research in humans are removed and if HGP's researchers are successful in extending certain techniques to humans. I assume for these cases that personhood can meaningfully be ascribed to embryonic human life after the possibility of twinning is past, which occurs sometime after implantation and about fourteen days of development.[7]

> *Case 1*: A young couple has just learned that they are both carriers of a recessive gene for Tay-Sachs disease, a recessive genetic disease that causes mental retardation, blindness, and early death. The presence of a single recessive gene for the disease does not produce symptoms in the couple, but if their two disease genes are passed to a child, it will invariably develop symptoms of the disease and die. This couple's children will have a 1 in 4 chance of inheriting these disease genes and a 1 in 2 chance of being carriers.
>
> Unfortunately, as many other couples who carry such genes, this couple only learned that they are carriers after their first child was born and diagnosed with Tay-Sachs disease. After three years of anguished helplessness, they watched the child die. Their doctors were sympathetic but could do very little. A year later this couple still grieve the loss of their first child, but they dearly want other children; moreover, they want children of their own, so they decide not to adopt and not to accept eggs or sperm from unaffected donors.
>
> Given these preferences, and given the current state of reproductive technology, they are informed when they go for genetic counseling that they have three options. First, they can proceed with a second pregnancy and simply hope that the two recessive disease genes will not be passed to their next child. Second, they can proceed with a pregnancy and test the fetus prenatally. Then, if the fetus is shown to be affected, they can either abort it or prepare for a second child who will again

> develop symptoms of Tay-Sachs disease and die. Or third, because of recent developments in molecular genetics, they can utilize an experimental procedure, the preimplantation diagnosis of genetic disease, to test embryos that have been fertilized *in vitro* for the presence of the disease genes *prior* to transfer to the woman's uterus. They decide to pursue the third option.

The preimplantation diagnosis of genetic disease essentially extends current *in vitro* fertilization techniques to fertile couples who are at risk of producing genetically affected children. The couple provides a number of eggs and some sperm to clinicians, who fertilize the eggs *in vitro*. Then, when the embryos—increasingly called "preembryos" to distinguish them from the implanted embryo as such—have developed for forty-eight to seventy-two hours, they can be biopsied and tested for the presence of certain genetic diseases. (In this case, "biopsied" means that one of the cells of the preembryo is removed for genetic testing; this can be done without harming the remaining cells of the preembryo, which will develop normally.)

> As a result of this process, clinicians now have four fertilized eggs from this couple to test. They discover that one of them is affected with Tay-Sachs disease and freeze it indefinitely (alternatively, they could simply discard it or use it for research purposes and then discard it). Of the remaining preembryos, one is normal and two are carriers. The couple elects to transfer the normal preembryo and the two carriers to the wife's uterus. One of the carrier embryos fails to implant and dies, but the other two develop normally. Nine months later, the woman delivers two healthy children, one of whom has no disease genes and one of whom is a carrier and, as such, will never develop symptoms of the disease.

The point is this: without such applications and the choices they make possible, *the particular children actually born to this couple would not have been born*. In this sense, then, the *existence*, *numbers*, and *identities* of these particular children are contingent on this application's development and use.

I now want to examine a possible future variation on the preimplantation diagnosis of genetic disease, represented by germ-line gene therapy.

> *Case 2*: Germ-line gene therapy will open a possible fourth option to this same couple, though to do so they will have to be willing to give birth to a child that is not entirely "theirs" from a genetic perspective. Let us suppose that this couple is not only opposed to abortion, but that they are also opposed to discarding or to freezing indefinitely their affected preembryo. If this couple were undergoing this procedure a few years from now, clinicians might be able to do "constructive genetic surgery" on the affected preembryo. That is, clinicians might be able to remove the affected DNA in the preembryo and replace it with DNA that has been engineered in the "proper" (that is, unaffected) sequence. We will assume that this couple does in fact elect this option, and again produces a healthy child.

This procedure is called germ-line gene therapy because the engineered genetic material is (ideally) reproduced and passed to all cells of the fetus as it develops prenatally, including its reproductive cells. If successful, this procedure would permit the child born as a result of it, and the future descendants of the child, to live without Tay-Sachs disease. The affected gene has simply been deleted from the child's germ-line (unless, of course, it is reintroduced by future matings with affected individuals). Again, however, because the re-engineered DNA is "foreign," the child who is actually born as a result of this procedure is *a different child from the one who would have been born without it*. The child has, in other words, an identity different from the identity it would have had without the intervention. So, if we assume that the enormous technical hurdles behind germ-line gene therapy are overcome, and that the particular preembryo in question survives transfer, gestation, and birth, germ-line gene therapy would also permit a person to come into existence who would not have existed otherwise. This child's existence and identity are, again, contingent on the development and use of this particular application.

The Problem of Contingent Future Persons

We can now ask why it is ethically significant that *different* future children are born using these interventions than would have been born without using them. Consider three types of action.

The first action is the type we are most likely to encounter in our everyday moral deliberations. It has *future effects on persons who are currently living*. For example, if a clinician deliberately gives a drug to a patient who does not need it or who is known to be allergic to it, the effects of this action are judged to be harmful. Or, conversely, when a clinician gives the same drug to a patient who needs it and who is not allergic to it, the effects of this action are judged to be beneficial. In more general terms, when we are trying to decide whether to attempt morally significant actions, the *foreseen effects of the actions on persons* generally count as reasons not to attempt or to attempt the action.[8] If the action in question can be predicted to leave these persons worse off or harmed in some way, this effect could count as a reason not to attempt the action, and if the action can be predicted to leave these same persons better off or benefited in some way, this could count as a reason to attempt it. The point is probably obvious: the fact that persons are affected by the action negatively or positively is one source of our commonsense moral reasons not to attempt or to attempt the action in question.

Second, some far-reaching actions are judged to have morally significant effects on *future* persons. When this is the case, we correctly reason in essentially the same way we did for the first type of action. Of course, the further we project the effects of our actions into the future, the more difficult it is to predict them. But insofar as our actions can be judged to hold significant benefits, costs, or risks for future people, we are rightly concerned with the effects of the actions on them. For instance, if we are trying to decide where to place a new waste dump, it will not be enough simply to locate it at a safe distance from present population centers. In one hundred years, population centers may shift, and people could live near the dump and be harmed. Thus, provisions need to be made to protect future people from such eventualities. Again, then, if the effects of such an action could leave future persons worse off or harmed, this could count as a reason not to attempt the action or as a reason to constrain it some way so as to minimize its negative effects.[9] The point of describing this second type of action is this: the fact that the future persons most affected by this action do not exist yet does not, in itself, give us a sufficient reason to ignore the possible effects of our actions on them. We might agree, for instance, that these effects at least give us a *prima facie* reason to pause and consider the interests or welfare of these future persons.

Now we must ask how these common-sense reasons fare when applied to a third type of action, illustrated above by the two cases from clinical genetics. This is a type of action *on which the existence, numbers, and identities* of future persons depend. Unfortunately, these reasons do not fare as well. Suppose we are considering an action like that described in Case 1, where we are using the newly developed technologies of molecular genetics to choose between genetically affected and unaffected preembryos. Assuming as we did in Case 1 that some preebryos survive transfer, implantation, gestation, and birth, we can reliably predict that healthy (that is, asymptomatic) children will be born. This is the outcome that both the parents and the clinicians desire; thus, they are clearly benefited by the action. But are the *children* themselves, born as a result of the intervention, better off? Have they benefited?

The answer is no, even though they are born free of a lethal genetic disease. It is true that these children will enjoy lives free of Tay-Sachs disease, but they have not *benefited* by the action, in the sense that they are better than they would have been had the action not been attempted; had the action not been attempted, *they* would not have been born. The reason they are not better off is that "better" appears to be a relational concept that requires us to make comparisons between a person's earlier and later states.[10] Such a comparison is simply not possible with this type of action; that is, we cannot meaningfully compare non-existence with existence. Thus, it makes no sense to argue that bringing these *particular* children into existence leaves them better off or benefited.

An analogous situation arises in Case 2, where we are trying to reconstruct a portion of DNA in order to avoid the moral ambiguities involved in discarding an affected preembryo. Is the healthy child born from such a procedure better off or benefited for having had its DNA re-engineered? Again, the answer is no. The child born as a result of this procedure is not the same child who would have been born had the procedure not be done. The re-engineered gene has altered the future identity of the person actually born as a result of the therapy, and other things being equal it will survive for many years free of Tay-Sachs disease.

But again, the particular child born from this procedure is not better off or benefited by the action; any other action would have meant that *it* would not exist.

Such outcomes may not seem ethically problematic to some readers, especially if we ignore here the very real questions concerning the moral and legal status of the preembryos that are discarded, frozen indefinitely, or used for research purposes. In the above two cases, the children actually born will live their lives free of genetic disease, and the parents and clinicians will realize the desired results of their medical intervention. However, an ethical problem can be imagined, for if we cannot benefit the children born from such choices, neither can we harm them.

Thus, for the sake of argument, let us alter these cases somewhat. Let us suppose that the couple in our above two cases decides to transfer a genetically *affected* preembryo (either selected through preimplantation diagnosis or created through germ-line gene therapy) to the wife's uterus. Granted, this is something they are *very* unlikely to do in the case of Tay-Sachs disease, but there may well be less extreme genetic diseases where parents want in fact to transfer genetically affected preembryos. For instance, such a choice may be elected by deaf parents who have reason to believe their children will be born hearing. Some deaf parents view deafness not as a disability but as an alternative life-style or culture, and thus may wish to transfer preembryos that will produce children that reflect their values (much as hearing parents might).[11] In the event such a choice were made, is the child born as a result worse off for having been born deaf? That is, has it been *harmed* by the action that brought it into existence? Again, the answer is no. This answer may seem counter-intuitive, but we must remember that the action simply does not leave the child who is actually born worse off or harmed as a result. Any other choice would mean that it would never exist. Of course, after we have an identifiable "person" in the deaf child (and this point will vary from conception to birth, based on one's substantive views of personhood), it can then be harmed or benefited just as any other person.[12] The problem of contingent future persons states only that the action on which a person's existence depends cannot be evaluated as a harm or benefit relative to that person—though, as we saw in the above cases, it is this action that holds so many life-long implications for the future persons actually born.

RESPONDING TO THE CHALLENGE

What should we make of this problem of contingent future persons? As we just saw, the problem apparently undermines common-sense reasons for undertaking actions that might benefit certain future people, while, at the same time, it undermines moral constraints on actions that might hold significant negative implications for the quality of future lives. We may conclude with a number of philosophers[13] that such reasons are the only valid grounds on which to base our ethics, and thus that these contingent future persons can make no moral claim on us; indeed, one libertarian legal scholar argues that the problem grants a sort of moral "carte blanche" for couples to pursue their "reproductive rights" without constraint.[14] But I find this conclusion hard to accept, for even if the couple who intentionally transfers a preembryo that leads to the birth of a deaf child cannot be said to harm that child, the child (and perhaps society in general) will bear the burdens of its deafness for life. Moreover, if we cannot benefit some future persons, there might be less reason to undertake long-range projects, and such projects have been the source of much that is good for human well-being. Thus, even though the logic of the problem of contingent future persons suggests otherwise, it would seem intuitively that these contingent future persons ought to make some claim on us. The "challenge" of the problem, however, is in discovering or in devising a way to justify such intuitions. This challenge proves *very* difficult to meet.

TWO CLASSES OF FUTURE PERSONS

The second half of the book is devoted to addressing the problem of contingent future persons. Here, though, I want merely to indicate an important preliminary implication of the discussion thus far and then summarize my argument. I believe the discussion suggests that we should make a distinction between two classes of future persons. We should distinguish between those future persons whose existence is *not* contingent on HGP's research in the relevant sense described above, and those future persons whose existence *is* contingent in the relevant sense. The former can be harmed or benefited by this research, but the latter cannot be harmed or benefited by it insofar as their existence, numbers, or identities are contingent on it.

This distinction will be crucial to an evaluation of HGP's implications for future generations, for the effects of HGP's research on the first class of future persons can apparently be addressed in terms traditional to bioethics (once, as I argue, the relevant moral domain is sufficiently expanded). However, the second class of future persons—those future persons whose existence, numbers, and identities are contingent on HGP's research—will require some other approach altogether; moreover, we will see that this other approach holds implications for addressing the first class of future persons, as our treatment of contingent future persons is logically *prior* to our treatment of non-contingent future persons. Given this basic distinction, let me summarize how the discussion will proceed.

Expanding the Moral Domain

My two basic aims are to investigate how HGP is likely to affect future generations and to ask what implications these effects have for evaluating HGP and far-reaching research efforts like it. These aims are complicated by at least two types of uncertainty. First, any attempt to predict outcomes for far-reaching policies such as HGP is fraught with empirical uncertainties that make it difficult and, in some cases, impossible to predict outcomes accurately. Second, the fact that the existence of some future persons will be contingent on HGP's research and applications introduces an element of ethical uncertainty into the mix as well; even if we can predict these effects accurately, we are not sure whether or how we can meaningfully evaluate them. To address these aims and their complications ethically, I believe we must expand the moral domain for thinking about consequences beyond that which is usually considered relevant in biomedical ethics, and we must do this in at least three ways.

First, we must go behind the clinical setting that has traditionally bracketed the moral domain of biomedical ethics and investigate the research *policies* that shape the problems faced by clinicians and their patients. This is the aim of chapter 2. A review of the congressional justification of HGP reveals how one such policy was shaped and justified. In this case, HGP's advocates appealed to four types of benefits:

biomedical, economic, political, and cultural. This appeal was supplemented by a cost-benefit analysis, though we will see that it essentially assumed the worthiness of these benefits and asked instead whether the proposed project was the most efficient means to reach them. The emphasis on efficiency proves to be very important for understanding the shape of the project and its likely implications for future generations.

The review of HGP's overall justification sets the stage for chapter 3, where I explore the likely effects of HGP's spin-off research and development efforts on future generations. This exploration suggests a second way to expand the moral domain. It implies that the reach of our responsibility must be pushed *temporally* much further into the future than is often the case in bioethics. ELSI's founders and other informed observers believe HGP's future applications will be developed roughly in three overlapping phases, with the second phase considered by far the most problematic. It is during the second phase that the above-mentioned information overload problems are expected to become most problematic and the therapeutic gap of clinical genetics will be most aggravated. We will examine these phases and their attendant problems in more detail in chapter 3. They raise an interesting question concerning the fairness of HGP's allocation of harms and benefits *between* generations.

Addressing this allocative question, however, must await an in depth exploration of the problem of contingent future persons, which I take up in chapters 4 and 5. In brief, I argue that the problem of contingent future persons requires us to expand the moral domain in yet a third way. Expanding it in the above two senses is primarily an exercise in consequentialist ethics, one that weighs for future generations the probable costs and harms expected from HGP's research and development efforts against its expected benefits. It is an empirically difficult exercise, and it is theoretically limited in ways that we will explore, but it is certainly within the methodological purview of traditional bioethics. This is not necessarily true of the third way we need to expand the moral domain.

Because an evaluation of HGP is complicated by the problem of contingent future persons, we must discover or devise a way to evaluate HGP's implications for *them*. This effort requires us to move conceptually from the consequentialist model used to address the above concerns to the *conceptions of value* undergirding this model. Indeed, I will argue that the problem forces us to choose between two fundamental approaches to value that are mutually exclusive and exhaustive. We must decide whether value is fundamentally "person-affecting," and thus attached to persons as such, or whether it is "impersonal," and thus attached to states of affairs more generally. However, this does not prove to be a straight-forward choice.

The problem of contingent future persons can be addressed by either approach to value, but certain ethical "costs" are attached to each, and we must decide which costs we can and cannot accept. A person-affecting approach to value enjoys a certain intuitive appeal because its lies behind many of our common-sense notions of harm and benefit. But it seems to exclude contingent future persons from the moral domain, because we cannot compare existence and non-existence in person-affecting terms (as we saw above). Thus, this approach to value can address the problem of contingent future persons only *indirectly* or by considering the probable effects of their existence or nonexistence on other "non-contingent" persons (that is, persons currently living and those future persons whose existence is not contingent on our actions in the relevant sense). However, I believe this approach restricts or narrows the moral domain in ways that leave future persons born as a result of the relevant actions overly vulnerable to the agents' interests. If we want to make them less vulnerable in ethical terms, we will need to *include* them in the moral domain. However, to include them requires an *impersonal* approach to value, and this approach may hold less intuitive appeal for many, especially in theological circles. An impersonal approach views value as fundamentally attached to the world, irrespective of the presence of "valuers" (persons). Interestingly, because this approach to value permits us to evaluate actions without regard to their effects on particular individuals, the existence, numbers, and identities of contingent future persons can be included *directly* in our moral deliberations. This can be done by reference to their probable contribution to the overall quality and quantity of life in the world, should they come into existence.

In chapter 4, I review two philosophical responses to the problem of contingent future persons that I judge to be the most sustained responses developed to date. Derek Parfit argues in *Reasons and Persons*[15] for the impersonal approach, whereas David Heyd argues, primarily against Parfit, in *Genethics: Moral Issues in the Creation of People*[16] for the person-affecting approach. My aim in this chapter is not to discuss exhaustively every nuance of their arguments, which would fill a book in itself, but to "map" the most important theoretical and practical implications of the problem, and to show where the most difficult problems lie. The purpose of this exercise is to set the stage for moving the discussion to theological grounds in chapter 5, which I view as the most constructive chapter of the book.

As I move to these theological considerations, I suggest that the problem of contingent future persons leaves theologically oriented ethicists—or, at least, *this* theologically oriented ethicist—in a possible dilemma. On the one hand, an impersonal approach to value offers ways to include contingent future persons directly in our moral considerations. It thus provides some of the moral constraints I—and, I suspect, other theologians will—seek for addressing the actions or policies on which the existence, numbers, and identities of future persons are contingent. However, adopting it may undermine theistic views of God altogether, as it implies that God serves an independent and impersonal scheme of value, rather than being the fundamental source or ground of value. On the other hand, a person-affecting approach permits us to preserve traditional views of God as the source or ground of value, though it seems to undermine the moral constraints I seek to address with regard to the problem of contingent future persons.

These concerns convince me to adopt a qualified person-affecting approach to value, the qualifications of which permit us to incorporate some of the constraints that are developed in Parfit's impersonal approach. I use three theologians to help me develop this qualified approach, namely, Richard A. McCormick, William P. George, and James M. Gustafson. McCormick's and George's Roman Catholic personalism provide a useful "foil" for developing a more impersonal position along the lines advocated by Gustafson. I believe it is possible to conceive of a theocentric approach to value, which I call an "impersonal theocentric" approach, that is formally person-affecting but qualified in ways so as to make it possible to include contingent future persons in the relevant moral domain and thus to consider them directly. My thesis is that an impersonal theocentric approach is better overall in addressing the problem of contingent future persons than the traditional theological approach to value, which I call "personal theocentrism."

In chapter 6, my conclusion, I attempt to draw the discussion together, asking what ethical significance the problem of contingent future persons holds for evaluating HGP's harms and benefits on future persons who are *not* contingent on it in the relevant senses. Here, yet another "cost" of an impersonal approach to value is uncovered, and it is one that my own impersonal theocentric position may not overcome. An impersonal approach holds troubling implications for evaluating the projected allocations of HGP's harms and benefits for future generations. That is, it may not give us any way to discuss the fairness (or lack thereof) of the allocation of harms and benefits between HGP's phase two generations and phase three generations. Nevertheless, the problem of contingent future persons must be seen as a logically prior problem when assessing HGP's implications for future generations whose existence is not contingent on HGP. This means that we cannot simply array the probable harms and benefits of HGP for future generations and evaluate them without first addressing this problem. And though I admit an attraction for impersonal approaches generally, finally I argue for a qualified person-affecting approach on theological grounds.

NOTES

1. Marc Lappé, "Long-Term Implications of Mapping & Sequencing the Human Genome: Ethical and Philosophical Implications," in *Mapping Our Genes: Federal Genome Projects: How Vast? How Fast?*, Contractor Reports, vol. 1, United States, Congress, Office of Technology Assessment (Springfield, VA: National Technical Information Service, February, 1988), p. 270 (p. 37 in original text).

2. Gregory Kavka, "The Futurity Problem," in *Obligations to Future Generations*, R. I. Sikora and Brian Barry, eds. (Philadelphia: Temple University Press, 1978), p. 192.

3. Victor A. McKusick, "Current Trends in Mapping Human Genes," *The FASEB Journal* 5 (January 1991): 12-20. McKusick compares the importance of this new paradigm to that of Vesalius's anatomical text of 1543, which formed the basis for William Harvey's (1628) physiology and Morgagni's (1761) morbid anatomy. Along these lines, he also suggests an anatomical metaphor may be more appropriate than the cartographic metaphor to describe the products of HGP's basic science.

4. United States, Department of Health and Human Services, Public Health Service, National Institutes of Health and United States, Department of Energy, Office of Energy Research, Office of Health and Environmental Research, *Understanding Our Genetic Inheritance; The U.S. Human Genome Project: The First*

Five Years, FY 1991-1995, DOE/ER-0452P (Washington, DC: National Institutes of Health, U.S. Department of Health and Human Services and U.S. Department of Energy, April 1990), Appendix 7, pp. 65-73.

5. By characterizing the problem this way, I do not intend to suggest that value assumptions are not embedded in the description.

6. The claim that identity is dependent on certain genetic interventions is not a claim that personal identity can be reduced to genetic make up (that is, that genetic make up is "sufficient" to explain identity), but merely a claim that personal identity is necessarily dependent on genetic make up.

7. The NIH and American College of Obstetricians and Gynecologists (ACOG) permit research on what is called "preembryonic" human life through the fourteenth day of *in vitro* existence and the development of the so-called "primitive streak" (though the NIH is not permitted to produce human embryos for research purposes). See ACOG Committee Opinion: Committee on Ethics, *International Journal of Gynecology and Obstetrics* 45 (1994): 291-301. Personhood cannot, of course, be predicated on scientific assumptions alone. For my own views, I follow Richard McCormick's arguments as outlined below in chapter 5.

8. I say "generally count" because an action may be morally indifferent, in the sense that leaves these persons neither worse off nor better off, or there may be competing reasons that override considerations of harm or benefit to the persons affected by the action. Here, however, I ignore these complications for the sake of clarity.

9. Again, we might discover competing reasons that would override these reasons. For example, as we try to project the effects of our action into the more distant future the problems of uncertainty become greater. Thus, we may argue that the empirical uncertainties are simply too difficult or in principle impossible to overcome. Or, the benefits to near-term future persons might be clearly demonstrable, and thus we might be tempted to "discount" foreseen harmful future effects on more distant future generations, arguing that future technologies will be developed to offset its negative implications. We will address some of these complications below.

10. I credit Professor Nicholas Fotion for this helpful clarification.

11. The possibility of selectively transferring preembryos that lead to the birth of deaf children was prompted by a conversation with some genetic counselors at a national conference, and confirmed as a likely prospect in conversation with Professor John C. Fletcher at the University of Virginia. The reader may find it hard to imagine parents choosing such lives for their children, but many geneticists believe it is only a matter of time before radical members of the deaf community and of the dwarf community request such services. It is an interesting case, because deafness and dwarfism are not uniformly seen as "diseases" or disabilities. In any case, we know for certain that some parents selectively destroy perfectly healthy preembyros, fetuses, and even children

simply on the basis of their sex preferences. Moreover, see Jerry E. Bishop's discussion of how this technology might used to select preembyros on basis of their genetic compatibility with cancer-stricken siblings in his, "Unnatural Selection," *Phi Kappa Phi Journal* (Spring 1993): 27-29. See also Kathy A. Fackelmann, "Beyond the Genome: The Ethics of DNA Testing," *Science News* 146 (November 5, 1994): 298-299.

12. This observation raises a related question that I will need to address below, concerning whether substantive views of personhood make a difference in how the problem of contingent future persons is defined. I have been assuming thus far that it is meaningful to talk of a "person" after the possibility of twinning is past. Below, however, we will see that the problem of contingent future persons can arise in other ways that do not require elaborate genetic technologies. Thus, I will argue that substantive views of personhood make a difference for what one can morally do to the preembryo, but not to the definition of the problem of contingent future persons. Under the relevant conditions, described more fully below, this problem can arise regardless of our substantive views on personhood or the beginning of persons.

13. See, for example, Thomas Schwartz, "Obligations to Posterity," in *Obligations to Future Generations*, R. I. Sikora and Brair Barry, eds., pp. 3-13.

14. This is John Robertson's argument in "Legal and Ethical Issues Arising with Preimplantation Human Embryos," *Archives of Pathology and Laboratory Medicine* 116 (April 1992): 430-435.

15. Derek Parfit, *Reasons and Persons* (New York: Oxford University Press, 1984).

16. David Heyd, *Genethics: Moral Issues in the Creation of People* (Berkeley: University of California Press, 1992).

2

The US Human Genome Project as Policy

> The key idea in genome projects was a dedicated effort to map and sequence whole organisms or significant parts of their genomes..."[1]
>
> Robert Mullan Cook-Deegan

> [T]he [congressional] debate was not about whether the research should go forward or not, but about how it should be managed and who should coordinate it. The march of human genetics would go on with or without a genome project, but resolution of management issues... would be more efficient in the long run if there were an administrative center or centers accountable to Congress.[2]
>
> James Dewey Watson and Robert Mullan Cook-Deegan

This chapter reviews the scientific and congressional justification for the US Human Genome Project as a basic research policy of the federal government. I will suggest that its "big science" shape—what Robert Cook-Deegan, an early analyst of the project for the Office of Technology Assessment, calls a "dedicated effort"—was the key policy issue confronting Congress when HGP was being debated. It also proves to be a key factor in shaping its likely three-phase future, outlined in chapter 3. Again, this brief review of policy is one way I believe we must expand the moral domain. Essentially, I am arguing that an ethical analysis of the likely effects of HGP's clinical applications on future generations should not be separated from the research policy that makes these applications possible. Said differently, the distinction between basic and applied research may be a useful analytical device for describing and predicting outcomes, but it should not be given too much weight when evaluating them.

SOME PRELIMINARY DEFINITIONS

The Human Genome Initiative (HGI) is a "shorthand for an international collection of scientific efforts to characterize the form and content of the human genome."[3] The Human Genome Project (HGP) is that portion of the world-wide Initiative being conducted in the United States and funded primarily by the US federal government through the National Institutes of Health (NIH) and the Department of Energy (DOE). The Howard Hughes Medical Institute (HHMI) has also contributed significantly from the private-sector.[4] The project began officially in this country with the start of fiscal year 1990 (1 October 1989) and is expected to take at least fifteen years to complete. Italy, the United Kingdom, France, the Commonwealth of Independent States, Japan, and Canada are involved in projects of their own and other countries are considering similar efforts. International coordination is being provided by the Human Genome Organization (HUGO).[5]

The fundamental scientific assumption behind this project is that structure determines function, that is, that the structure of a molecule determines its physical shape which in turn determines its function. By knowing the structure of a molecule, researchers can (better) predict its function. Though reductive, this assumption has proven to be highly predictive in earlier research.[6] The molecules whose structure researchers seek to determine in this project are known collectively as the "genome."

The term "genome" is used to describe the genetic make-up of both individuals and species. The "human genome" is defined as all the genetic material, that is, all the deoxyribonucleic acid (DNA), contained in the chromosomes of each and every human cell that has a nucleus (except egg and sperm cells which have only half as much), plus the genetic material in the extra-nuclear mitochondria (which is inherited from the mother). Each human chromosome contains two copies of a huge DNA molecule which are almost (but not exactly) identical. DNA is a macromolecule composed of four nucleotide bases, arranged in the well-known double-helix form that was first discovered in 1953 by James Watson and Francis Crick.[7] Its four bases include two purine bases, adenosine (A) and guanine

(G), and two pyrimidine bases, cytosine (C), and thymidine (T). In nature, guanine pairs only with cytosine, and adenosine only with thymidine. Hence, the term "base-pairs": by determining the order of one strand of DNA the other can be deduced. There are thought to be approximately three billion nucleotide base-pairs in the human genome, with roughly three million base-pair differences (0.1%) between any two individuals.[8] HGP's goal is to characterize a *single, composite* human genome which, for HGP's purposes, is defined as twenty-two (22) autosomes, an X and Y chromosome, and an extra-nuclear mitochondria.[9]

Genes may be thought of as concentrated nodes of DNA, distributed along the chromosomes, that serve as the basic physical and functional units of heredity. Genes "code" for specific amino acids, which are strung together to form peptide chains and finally, from these chains, complex proteins. There are thought to be anywhere from 50,000 to 100,000 genes in the human genome, and the major "structural" or "housekeeping" genes are thought to be in the same locations for all individuals.[10]

The structure of the genome will be represented by various types of "maps." In general terms, gene mapping is defined as the "assignment of genes to chromosomes,"[11] though mapping techniques are now considerably more refined than this definition suggests. There are two basic types of maps, genetic linkage maps and physical maps, and they are distinguished both by their units of measurement and by their levels of resolution.

Genetic linkage maps locate genes statistically, being "derived from meiotic recombination frequencies and measured in centimorgans (cMs)."[12] Compared to certain physical maps, linkage maps have a relatively low resolution. Since the 1970s, however, more precise linkage maps have been created using restriction fragment length polymorphisms (RFLPs), which "mark" inherited base changes with restriction enzymes. All linkage maps are in principle limited as clinical diagnostic tools, however, because they require access to and cooperation of extended families for their construction. Nevertheless, they are useful as research tools in targeting general areas on a chromosome to investigate for specific genes and, in this capacity, they have been extremely helpful in isolating many of the major disease genes located thus far by researchers.

Physical maps, as their name suggests, measure the physical distances between various sites on the DNA. Low-resolution physical mapping technologies include somatic cell hybridization, chromosome sorting, karyotyping, chromosome banding, and *in situ* hybridization. High-resolution physical mapping technologies use cloned "vectors" (stable, known DNA fragments cloned into viruses) to produce overlapping segments of DNA which can then be ordered on the basis of their overlapping segments (the so-called "contig" maps). At its highest resolution, a physical map of the human genome will provide researchers with the actual base-pair sequence of an entire genome, a DNA molecule, or a relevant portion thereof.[13]

It is the sequence of the genome that HGP researchers ultimately seek as it will provide them with the most precise predictions about molecular function, but on the way to this goal they will produce high resolution maps of both types. Indeed, HGP's fifteen-year goal is to provide researchers with what is variously called a "library," an "encyclopedia," or a "code book" of maps of the entire human genome and, additionally, of the genomes of several other model organisms (that is, other animal models). At the recommendation of the National Research Council (NRC),[14] the first half of the project is devoted to producing high-resolution linkage maps and the second half will concentrate on sequencing, after (it is hoped) sequencing technologies are available to speed up the process and to lower its costs. It is assumed that by knowing the locations of genes and of the relevant nucleotide sequences, gene function (and thus malfunction) can be more easily inferred. The model organisms will permit researchers to confirm experimentally the function of specific genes by tracing gene products on the chemical level. Speaking metaphorically again, these organisms will help researchers decipher the internal grammar of the human genome's code book.

Two Interpretations of HGP

Described in this way, HGP is probably best viewed as a basic research or science project whose principal product, like other similar efforts, is new knowledge or new information.[15] And, indeed, if one reads the literature or listens to the scientists who are actually involved with HGP research, one gets the impression that this is the dominant view with which they operate. However, this is not the only way to view HGP, and it is doubtful that the project would have been funded by Congress if it were. In fact, the project was justified almost from its initial conception by the applied research or technology expected to be spun off from it or made possible on the basis of it. Applied research is aimed at turning the results of basic research into commercially useful products.[16]

Thus, it is with HGP's expected applications in view that the project is also interpreted as a great national undertaking and compared to the Apollo Moon Program (and, less frequently, to the Manhattan Project).

> Similar to the 1961 decision made by President John F. Kennedy to send a man to the moon, the United States has committed itself to a highly visible and important goal [of mapping and sequencing the human genome]...A more important set of instruction books will never be found by human beings.[17]

Of its many anticipated applications, HGP is expected to lead primarily to biomedical technologies. These in turn are expected to permit clinicians routinely to diagnose, prevent, treat, and eventually cure more than 3000 known genetic diseases and those multifactorial diseases in which genetic predisposition plays a role.

> When finally interpreted, the genetic messages encoded within our DNA will...not only help us understand how we function as healthy human beings, but will also explain, at the chemical level, the role of genetic factors in a multitude of diseases, such as cancer, Alzheimer's disease, and schizophrenia, that diminish the individual lives of so many millions of people.[18]

Moreover, the new biotechnologies that will permit these medical feats are expected to play a significant economic role by helping the fledgling US biotechnology industry maintain its tenuous lead in the world, with applications expected in such diverse fields as agriculture and waste management.

Two Levels of Analysis

We will see that HGP will affect future generations in both of these senses, that is, both as basic research policy and through the applications it spins off. However, to the extent that these effects on future generations can be discerned and anticipated, they are ethically relevant for agents of the current generation primarily in *justifying* such policies and applications. In this book, I will treat the justification of policy on two levels: first, the justification of our *collective* actions as a society, that is, the justification of our public policies by collectivities of agents (represented in the case of HGP by the United States Congress); and, second, the justification of *individual* actions as particular citizens choose to use or not to use the products of our public policies (represented in the case HGP primarily by the couples and clinicians who use its applications).

This distinction between the collective and individual levels will be important throughout the book because of how it influences our justifying arguments. The distinction consists both in the number of agents involved in any given choice relative to an action or policy under consideration, and in the scope of the relevant outcomes that must be considered when predicting and evaluating consequences. Put simply, the outcomes of choices made on the collective level are much harder to predict, but will potentially affect many more people and, in some cases, for many years to come. We must thus consider a moral domain with a much wider scope on this level. This is not to suggest that the use of HGP's applications by individuals will not also require public policy regulation and justification, but it is to suggest that we cannot simply consider these individual choices in isolation from the collective choices that make them possible.

In the remainder of this chapter, then, I will review and summarize the arguments that were used by Congress on the collective level to authorize HGP's funding as biomedical research policy. We will explore below what implications this policy choice might hold for future generations.

OF THRESHOLDS AND CATALYSTS

HGP was funded directly by Congress through special authorizations and appropriations. However, while its three billion dollar price tag makes HGP the largest single biological research project in human history, relatively speaking this is not a particularly large sum of money, and Congress would not ordinarily concern itself with such amounts. Thus, there are two initial questions that need answers: why (in a descriptive sense) did Congress concern itself with the Human Genome Project and why, then, did it fund HGP? That is, to what reasons did Congress appeal to justify its decision?

GETTING ON THE CONGRESSIONAL AGENDA

The steps that lead to the authorization of and appropriation for a given congressional policy are extremely complex in empirical terms.[19] However, for our purposes, these steps can be abstractly conceptualized as a two-step process that involves, first, getting an item on the congressional agenda and, second, having that item selected over competing items or over competing versions of the same item. In practice, the justification of an item on the congressional agenda assumes the legitimacy of the (ostensibly) democratic processes that put it there, though in theory the two steps are of course mutually dependent. For purposes of this discussion, I am concerned primarily with the selection of one item on the agenda over another, which is to say I do not intend to question here the legitimacy of the processes that put an item on the agenda.[20] I will, however, try to summarize the reasons that were offered to move HGP onto the congressional agenda and to select it over alternative possibilities.

In short, HGP was put on the congressional agenda in the mid-1980s, first, as a way to settle an inter-agency "turf battle" concerning which agency should serve as the "lead agency" for the project and, second, as a response to a number of scientific, economic, political, and cultural factors that proved to be influential at the time. When HGP was finally authorized and funded from among competing alternatives, Congress also appealed to a rather shortrange cost-benefit analysis to support its choice.

The Key Concept

As it was originally conceived and proposed, HGP was not intended to begin something new *scientifically*; rather, it was intended to coordinate under a new *administrative* structure what had already begun in small and "piecemeal" ways under the funding and research practices traditional to biological and biomedical investigation. This distinction is important, at least from the perspective of the advocates of HGP, for they did not claim that they were proposing something revolutionary in scientific terms. Rather, what they consider revolutionary about HGP is its centralized or coordinated organizational structure and, moreover, its public commitment *as a basic research project* to reach certain goals within specific time and budget constraints.[21] This was new in biology, and for this reason HGP is often referred to as the first "big science" project in the field. Such an approach is described by Robert Cook-Deegan, a physician then with the Office of Technology Assessment (OTA) and in charge of an influential 1988 report to Congress on the proposed project, as a "dedicated effort."[22] This is the key concept behind HGP, and it was the concept around which most of the initial controversy concerning HGP's funding coalesced.

A Response to Agency In-Fighting

The idea of a "dedicated" genome project occurred at almost the same time, though independently, to at least three different (and rather entrepreneurial) researchers. The first proposal came from Robert Sinsheimer, then Chancellor of the University of California at Santa Cruz, at a meeting held in May 1985.[23] He was casting around for ways to invest a large financial gift that had been given to the University, and he initially wanted to use it to build a genome research center. However, as large as this gift was, he needed even more money to begin such a center and these additional funds were not forthcoming. He did, however, manage to plant the idea of a such project with a number of other researchers, most notably Walter Gilbert, who later argued supportively for it with a number of their more reluctant colleagues.[24] (The original gift ended up in the Keck telescope on Mauna Kea mountain in Hawaii.)

A second and more successful source was Charles DeLisi, then head of the Office for Health and Environmental Research at the Department of Energy (DOE). In keeping with the DOE's long-standing responsibility to track mutations resulting from the atomic bombs dropped on Japan at the end of World War II, DeLisi was looking for better ways to measure heritable mutations in humans. He believed that sequencing the entire genome would advance this study, and he was successful in securing some uncommitted funds to begin a pilot project at a DOE laboratory.[25]

However, the DOE's decision to move ahead with its project caused a reaction within another federal agency, the National Institutes of Health (NIH). The NIH funds most of the basic biological and biomedical research done in the US, and a number of prominent researchers working at or funded through the NIH argued that, if such a project were to be undertaken at all, the NIH should be the "lead agency." The reasons cited for their position concerned the DOE's poor track record with biological research and the fact that tracking mutations was simply not an adequate scientific justification for undertaking such an ambitious project. Younger researchers funded through the NIH were also concerned with protecting their own funding sources, and they argued that the NIH should stay clear of the massive project altogether unless "new monies" could be found.[26] At that time, however, the NIH had no available funds to commit to the project and securing new funds would mean approaching Congress directly. Congress had been pressuring the NIH for some years before this to constrain its research budget, and thus NIH officials were initially reluctant to make such an approach.[27]

In the meantime, Renato Dulbecco, a Nobel laureate and then President of the Salk Institute, was looking for more efficient ways to advance cancer research. He argued persuasively that genome research would be the best way to proceed in the future fight against cancer,[28] and cancer research had long been supported by Congress through the NIH. Dulbecco's proposal, and others like it, were eventually utilized by NIH-sponsored researchers to argue that the project should indeed go forward, that it should be aimed more directly at *biomedical applications* and, therefore, that it should be funded through the NIH. The DOE, it was argued, simply did not and would not realize the potential of such an undertaking. DOE advocates debated this charge, of course, and eventually the argument caught the attention of Congress. This agency in-fighting thus put HGP on the congressional agenda. In the process, however, the in-fighting also helped to clarify the goals that scientists thought HGP ought to pursue, and these goals placed it more squarely in the recognized purview of the NIH.

A Response to Opportunity

This said, HGP was not funded simply to settle the internal wrangling of competing federal agencies. (In any case, just before Congress was scheduled to decide the issue of the lead agency, the DOE and the NIH signed an unprecedented joint agreement to cooperate on the project and thereby avoided an imposed solution.)[29] When the proposed project was finally placed before congressional leaders they proved to be enthusiastic about it for their own reasons. HGP received serious consideration because there was a history of successful biomedical research behind it to which Congress could respond and also because the time seemed right (to Congress) to mount a major new effort in biomedicine.

The successful research behind HGP emerged in the three decades prior to the project's founding, as powerful tools developed in the field of molecular biology migrated into (Cooke-Deegan says "invaded") the field of human genetics.[30] This gave rise to a new field called molecular genetics or, as it is more popularly called, the "new genetics."[31]

Molecular biology was essentially founded in 1953, following the discovery of the structure of DNA by Watson and Crick. Over the next twenty years, molecular biologists were remarkably successful in developing techniques and procedures for investigating the structure and function of genetic material, beginning first with simple and then moving to ever more complex organisms. By 1973 researchers had developed recombinant capabilities and this soon led to the ability to clone large segments of DNA in a single experiment. In the late 1970s and early 1980s, a number of other tools were developed, including some that have already been mentioned.[32] These tools included restriction fragment length polymorphisms (RFLPs), a tool for constructing high-resolution linkage maps;[33] somatic cell hybridization, *in situ* hybridization, and chromosome sorting, all examples of low-resolution physical mapping techniques;[34] pulsed-field gel electrophoresis (PAGE), which separates DNA fragments of 100 to 1000 kilo-bases;[35] yeast cloning with yeast artificial chromosomes (YACs), which will clone sequences of specific sites of DNA;[36] and, perhaps most importantly, polymerase chain reaction (PCR), which permits rapid amplification of short DNA segments.[37]

However, it was as these developments in basic molecular genetic research began to make their way into *clinical applications* that researchers became more hopeful about the possibility of routinely diagnosing, preventing, and treating genetic disease in humans.[38] Before these developments, human geneticists were (and, for the most part, still are) limited mainly to studying the rates of disease transmission and to the clinical description of disease progression. There is little they can do to treat individuals who suffer from genetic diseases and disorders, and prevention in the field is too often reduced to "therapeutic" abortion. Moreover, the percentage of patients with genetic-related problems has increased dramatically in developed countries of the world, due largely to the success of environmental medicine in decreasing other types of disease. In these developed countries, it is currently estimated that as many as one third to one half of all pediatric admissions to hospitals are related in some way to genetic problems, and genetic disease is a leading cause of death among pediatric patients. The few treatments that have been developed are often excruciatingly painful, must be spread over a lifetime, are often not very effective, and are exceedingly costly. If present rates of expansion continue, genetic disease could soon consume the entire health care budgets of some countries.[39]

By the mid-1980s, however, researchers began to argue that the new developments in molecular genetics offered hope that the long-standing clinical impasse in genetics (its so-called "therapeutic gap") could and should be overcome. Dulbecco's suggestion that cancer research be advanced along these lines was merely the first of many such arguments. Contributing to this growing enthusiasm was the isolation of genes related to about 100 genetic diseases.[40] So, with the emergence and growing promise of molecular genetics, the biomedical research community found itself at a threshold of sorts.[41] The notion of "routine" or "standard" medical use is the key to understanding this particular threshold. Advances had brought researchers to a point where many were claiming it was feasible to begin developing genetically-based screens, tests, and therapies that could be used *routinely* in "mainstream medicine" to detect, prevent, treat, and eventually cure the thousands of known genetic disorders and diseases, as well as those multifactorial diseases in which genetic predisposition plays a role. It was also claimed that surmounting this threshold would yield no less than a "new paradigm"[42] for all of scientific medicine, a reductive paradigm that would permit researchers to trace the root biological causes of disease to the molecular level and open new approaches to understanding human development and evolution.

In truth, however, this utopic vision was only arguably feasible. Indeed, the magnitude or the scale of the task required of researchers to bring these applications into routine medical use taxes the imagination. Assuming that each base-pair of the human genome's three *billion* base-pairs corresponds to a character, the code book of a single composite human genome will be equivalent to approximately thirteen complete *sets* of the *Encyclopedia Britannica*.[43] Moreover, without a grammar to decipher this code, researchers will have no idea what much of it "means" (that is, how it functions). We must imagine thirteen complete sets of encyclopedias without spaces, periods, paragraphs, or indexes—simply three billion characters, one after the other, with no way to know where one functional unit (that is, one gene) begins or ends.

Researchers are actually further along in their understanding than this description suggests, but perhaps not much further. Indeed, James Watson testified that it could take researchers another one thousand years to decipher or interpret DNA function.[44] Watson has been accused of hyperbole in this prediction, but it cannot be dismissed entirely. If his estimate is accurate, say, only by a quarter or even a tenth, that is still 100 to 250 years of concentrated effort needed simply to interpret HGP's fifteen years of data collection. (I will return to these time estimates below when we consider more directly the impact of this research policy on future generations.)

In any case, researchers were increasingly aware that in order to realize the new biomedical capabilities envisioned by clinical geneticists, they would require much more basic knowledge of the structure and function of the human genome. This reality formed the basis of the principal justification for the "big science" shape of the "dedicated" project. The "piecemeal" or "cottage-industry" approach traditionally used in biological research would simply not suffice, they argued, to take researchers through this particular threshold *in an acceptable time-frame*.[45] The dedicated US Human Genome Project is thought to be a rational response to this threshold problem and, as such, I believe we can best understand it as a "catalyst" for a massive social experiment.[46]

In chemical reactions, the presence of a catalyst lowers the energy threshold (and thus the time) required for a given reaction to begin. In effect, the organizers of the Human Genome Project argued that HGP would serve as a catalyst for the research and development of new biomedical applications. Without something like HGP, they argued, the piecemeal approach traditionally used in biological and biomedical research would simply not suffice to surmount the threshold between the present "state of the art" in clinical genetics and the improved tests, treatments, and cures envisioned by researchers. Advocates of HGP conceded that the project would increase short-term costs, but they argued that it would lower overall long-term costs—again, *when compared to traditional approaches to biological research*. HGP can be viewed as a social experiment—perhaps, indeed, like the Apollo Moon Program or the Manhattan Project—because we do not yet know whether it will ultimately be successful in reaching its ambitious and publicly stated goals within the time-frame and budget set for it or, if successful, whether society will be able to adjust in the wake of its products.

This said, other factors also influenced Congress, chief among them certain economic considerations.[47] Economic benefits are expected of HGP as its emerging biotechnologies are "transferred" to private and public sector laboratories for development and commercial exploitation. Behind this economic argument is a larger story of the Cold War's ending that I can mention only briefly.

HGP was being debated as the Berlin Wall was coming down and the Eastern Block was breaking up. In this context, Congress was trying to decide how to shift the US economy from its defense orientation in an orderly way, and part of these deliberations included an extended debate on how to increase US competitiveness in the global market. Congress was persuaded that biotechnology is one area in which the United States retains a competitive "edge" over the Japanese and the then European Community (EC). However, it was (and is) commonly agreed that the US needed to increase the efficiency of its technology transfer processes if it were not to lose this edge. Again, HGP's "dedicated" shape was thought to be a rational response to this concern. By taking advantage of our tradition of strong investment in basic science, HGP was perceived as one way to propel the fledgling US biotechnology industry ahead of its nearest competitors.

The economic argument behind HGP may also help illumine some of the political motivations behind it. In lieu of an appeal to the Cold War, the economic benefits expected of HGP provided politicians with a certain "legitimation potential" when allocating funds. For example,

Senator Pete Domenici asked in a 1987 hearing concerning HGP, "What happens if peace breaks out?"[48] In effect, he was asking how Congress, without the threat of the Cold War, could legitimate funding for defense-oriented laboratories like those the DOE operates in Los Alamos, New Mexico, Domenici's home state. To meet this challenge another "war" was proposed, and the language of the Cold War was simply transferred by some congressional supporters of HGP to the "economic war" being waged with the EC and, especially, with countries on the Pacific Rim.

In light of these expected economic and political benefits, it is worth noting that HGP may also have been funded publicly in order to *limit* the market's influence on this potentially threatening research. That is, HGP may have been funded not only to produce certain benefits or goods, but also to *avoid* certain harms or evils. Because this research is directed at the *human* genome, it engendered strong, almost visceral reactions in some segments of the American public. Often, for instance, the Frankenstein metaphor was used to warn of the intrinsic evils of "tampering" with the human genome.[49] These kinds of reactions suggested that this research, if allowed to develop in certain ways, could create a significant political backlash for policy makers. However, by allocating funds for HGP, Congress gave itself a way of directing the research largely toward biomedical ends without directly mandating these ends. In effect, then, Congress created an *incentive* for researchers to pursue research and to develop applications that it viewed as socially desirable.[50] Thus, Congress was not only responding to opportunities presented to them by scientists, but also actively nudging this potentially threatening research into socially beneficial channels.[51]

Finally, the project is also expected to yield certain cultural benefits, and these also served to motivate at least some congressional leaders. These cultural benefits are among the most nebulous and least quantifiable expected of HGP. They include a certain "political capital" that nation-states are thought to enjoy in the wake of significant scientific achievements.[52] They also include what might be called (for want of a better term) "bragging rights" for having done something that is presumably good for future generations, and for having done it first. This suggests that there was at least some awareness of future generations when the project was funded, and it seems to assume that future generations will, in fact, benefit from HGP's research. However, before we address questions about the relation of HGP to future generations, we must investigate the story behind HGP's funding somewhat further. For getting the project on the congressional agenda and debated was only part of the process of congressional justification.

FROM THE CONGRESSIONAL AGENDA TO POLICY

It was clear from the start that Congress was intrigued with HGP as a concept. Moreover, it was clear that certain congressional supporters of it, such as Senator Domenici, would fund it in any case for politically expedient reasons. But Congress also wanted to know if the proposed project would in fact be the most efficient means to the ends envisioned for the research. This question considerably narrowed the debate surrounding the project. In the end, Congress had essentially two alternatives before it. It could decide to fund the proposed project, or to do nothing and by default permit genetic research to continue in the traditional way. Congress thus assumed, or was sufficiently persuaded, that the ends this project is designed to pursue were worthy ends, and that they would be realized by traditional research methods in any case, just much more slowly. Also, it did not ask whether there was some better way overall to advance human health. Indeed, it is generally conceded that by the time Congress received its 1988 report from the Office of Technology Assessment (OTA) the project was going to go forward in some form—in fact, some funds were appropriated for it even before this report was received.[53] The only question remaining to decide was whether HGP was in fact the most efficient means to reach these biomedical ends.

Comparing Costs and Benefits

When Congress finally decided this question of means, it did so by projecting and comparing the probable costs and benefits of the various alternatives before it. To do this, it used a straightforward cost-benefit analysis. Though much criticized, the use of cost-benefit analysis continues to expand on all levels of the federal government, and with state and local governments and large corporations as well. It was first introduced to the federal bureaucracy during the New Deal and its use expanded until, in the 1970s, President Ford required by executive order all departments of the federal government to review their proposed regulations in cost-benefit terms. This executive order has been renewed under each administration since Ford.[54] Congress gained direct access to analysts trained both in cost-benefit analysis and in related methodologies[55] when it created the Office of Technology Assessment (OTA) in 1972. The OTA was created to free Congress from its dependence on the executive branch's policy experts. It does not make recommendations directly to Congress, but it does help Congress analyze, anticipate, and plan for the consequences of technological change and development.

Why does Congress use cost-benefit analysis? It is primarily an economic tool and perhaps its appeal can be best explained in these terms. The growth of the federal government in this century has resulted in the increasing interpenetration of the political and economic spheres.[56] Given this increasing interpenetration, the need for something like a cost-benefit analysis is rather easily understood, at least on an intuitive level. This need can be illustrated by contrasting the production of private goods with the production of public goods.[57]

Assuming a market system that exists under conditions of relative scarcity, the production and exchange of private goods is governed largely by their price. The price of a good is one way to represent the value of a particular good for both the seller and the buyer. For the seller, it is an expression of the costs and risks involved in bringing the good to market, and for the buyer it is an expression of its desirability. In this way, then, the price system provides both the seller and the buyer with important information about optimal levels of inputs and outputs, and thus functions as an efficient *feedback* mechanism. However, with the production of public goods, such as those represented by the new knowledge of HGP's basic science, it is precisely this feedback mechanism that is missing. Government policies have implications for the economy and for society in general, but government policy makers have a much harder time evaluating the optimal level of system inputs and outputs than their private-sector counterparts. This introduces a considerably greater level of uncertainty into the decision-making process for governments.[58]

Cost-benefit analysis is one method, then, that government decision-makers use to determine an optimal input-output level, and it is thus one way to reduce some of the empirical uncertainty related to their decision-making. The primary objective of cost-benefit analysis is "allocative efficiency."[59] Formally, this is what it means to say a policy is "justified" in cost-benefit terms. Allocative efficiency means the policy alternative chosen is the most efficient means available to reach the intended ends. As such, cost-benefit analysis may be defined as

> a conceptual structure and set of techniques for relating means to ends, for arranging the various costs associated with each course of action, and for describing, comparing, and assessing possible outcomes.[60]

It is an example of what is known more broadly as end-state analysis.[61]

For a basic research project like HGP, the always difficult cost-benefit analysis is even more problematic. Basic research is inherently unpredictable and predicting which applications might result from certain basic research efforts may not be much better than an educated guess.[62] Also, any project which has goals that are intended to be realized in the future may be interrupted by intervening agents, making it unpredictable in principle. These kinds of considerations helped to convince the advocates of HGP that it would be politically expedient from a funding perspective to produce practical (applied) results very quickly, which they have done. In any case, HGP is estimated to cost about three billion dollars, spread over fifteen years at approximately two hundred million per year. These costs represent the project's direct *opportunity* costs, the definition of which also is rooted in the problem of economic scarcity.[63]

Because of scarcity, the costs of pursuing HGP's goals means that other, perhaps equally worthy projects cannot be pursued. The foregone benefits of these other projects are called opportunity costs. But opportunity costs can also refer to the projected costs of *not* pursuing the benefits of the particular policy under consideration, that is, the costs of its foregone benefits. Thus, for example, suppose the US decided not to pursue HGP's biotechnical benefits and the Japanese did pursue them. The Japanese might eventually sell the products of this research back to us at many times the cost of developing them ourselves. This fear may help explain why technology transfer has become such a concern with this project. Most observers are not worried that the Japanese will jump ahead of the US on basic biological research, but they are worried they will take the findings of US basic research and develop biotechnical applications more quickly. Again, these considerations became supporting reasons for undertaking a "big science" or "dedicated" project.

HGP's Projected Costs

The two hundred million dollar per year figure was first calculated by the National Research Council (NRC), and divided into three main categories: personnel ($120 million per year for 1,200 researchers at $100,000 per researcher per year), construction and equipment ($55 million per year), and administrative and coordination costs ($25 million per year).[64] The OTA's 1988 report largely confirmed the NRC's projections.[65] At the time these reports were produced, the direct cost of mapping a base pair sequence was between two and five dollars, depending on the experience of the laboratory in question. To have a realistic hope of meeting its ambitious goals, HGP researchers estimated that they needed to lower this cost to approximately $.50 per sequence.[66]

There is no reason to doubt these figures, but the reader should note that HGP's cost projections were so sensitive to rapidly developing technologies that policy analysts would not guarantee them beyond two years. Moreover, due to the same rapidly developing technologies, *no* estimate could be given for the indirect costs of the project; nor, of course, could the costs of the spin-off projects be estimated. These spin-off projects will pursue applications using the new knowledge gained through HGP and, as we learned above, they were essential in justifying the project.

As inadequate as these cost projections are, I believe they were nevertheless persuasive for Congress on the basis of a consideration of the alternatives before it. Consider, again, the choice Congress faced. It needed to decide whether the new approach represented by HGP, at approximately three billion dollars spread over fifteen years, was better (that is, more efficient) than the traditional research approach to genetic research, which was estimated to take between six and fifteen billion dollars spread over the next *one thousand years* (estimated on the basis of the then-current rate of sequencing, roughly two million base-pairs per year at two to five dollars per base-pair).[67]

Faced with these alternatives, HGP did not present itself as a "hard choice" to Congress. Again, the threshold and catalyst metaphors are useful in characterizing this situation. HGP is assuredly a massive undertaking. But in mapping and sequencing the human genome, researchers are trying to do what is merely(!) a first and purportedly necessary step in basic science toward realizing the more socially useful biotechnology products expected of HGP. By giving the project a "dedicated" or "big science" shape, Congress intended to take this first as efficiently as possible. That is, Congress wanted to save time—time, in this case, representing both money and, finally, lives saved—and time will eventually prove Congress right or wrong in its choice.

NOTES

1. Robert Mullan Cook-Deegan, "The Human Genome Project: The Formation of Federal Policies in the United States, 1986-1990," in *Biomedical Politics*, ed. K. E. Hanna (Washington, DC: National Academy Press, 1991), p 108.

2. James Dewey Watson and Robert Mullan Cook-Deegan, "Origins of the Human Genome Project," *The FASEB Journal* 5, no. 1, (January, 1991): 10.

3. James D. Watson and Eric T. Juengst, "Doing Science in the Real World: The Role of Ethics, Law, and the Social Sciences in the Human Genome Project," foreword to *Gene Mapping: Using Law and Ethics as Guides*, eds. George J. Annas and Sherman Elias (New York: Oxford University Press, 1992), p. xv.

4. HHMI has played a key role in funding certain aspects of the project that would have been problematic for the federal government to fund, especially those efforts to create international linkages among researchers. See Watson and Cook-Deegan, "Origins of the Human Genome Project."

5. Victor A. McKusick, "The Human Genome Organization: History, Purposes, Membership," *Genomics* 5 (1989): 385-387. The United Nations Educational, Scientific, and Cultural Organization (UNESCO) has also involved itself in an effort to insure the benefits of this research are not restricted to developed countries alone.

6. But compare Evelyne Shuster, "Determinism and Reductionism: A Greater Threat Because of the Human Genome Project?" in *Gene Mapping: Using Law and Ethics as Guides*, eds. George J. Annas and Sherman Elias (New York: Oxford University Press, 1992), pp. 115-127.

7. J[ames] D. Watson and F[rancis] H. C. Crick, "Molecular Structure of Nucleic Acids: A Structure for Deoxyribose Nucleic Acid," *Nature* 171 (25 April 1953) 737-738.

8. Mary Ann G. Cutter, et al., *Mapping and Sequencing the Human Genome: Science, Ethics, and Public Policy* (Colorado Springs, CO: BSCS and American Medical Association, 1992), p. 10.

9. Ibid., p. 1.

10. Ibid.

11. United States, Congress, Office of Technology Assessment, *Mapping our Genes: Genome Projects: How Big, How Fast?*, OTA-BA-373 (Washington, DC: U.S. Government Printing Office, 1988), p. 26.

12. Victor A. McKusick, "The Human Genome Project: Plans, Status, and Applications in Biology and Medicine," in *Gene Mapping: Using Law and Ethics as Guides*, eds. George J. Annas and Sherman Elias (New York: Oxford University Press, 1992), p. 19.

13. United States, Congress, Office of Technology Assessment, *Mapping our Genes: Genome Projects: How Big, How Fast?*, OTA-BA-373, pp. 26-44.

14. The NRC argued that researchers should "map first, sequence second." National Research Council, Commission on Life Sciences, Board of Basic Biology, Committee on Mapping and Sequencing the Human Genome, *Mapping and Sequencing the Human Genome* (Washington, DC: National Academy Press, 1988), p. 2.

15. Peter H. Aranson, *The Political Economy of Science and Technology Policy*, unpublished manuscript (Atlanta, GA: Emory University, 1988), pp. 6-7.

16. Ibid.

17. James D. Watson, "The Human Genome Project: Past, Present, and Future," *Science* 248 (6 April 1990): 44.

18. Ibid.

19. For a helpful description of these complex processes that utilizes an interactional model, see Randall B. Ripley and Grace A. Franklin, *Congress, the Bureaucracy, and Public Policy*, 5th ed. (Pacific Grove, CA: Brooks/Cole Publishing Company, 1991).

20. The broader question of legitimacy takes us too far afield into democratic theory. On this question, however, I am informed by Robert A. Dahl, *Democracy and Its Critics* (New Haven: Yale University Press, 1989).

21. See Watson, "The Human Genome Project: Past, Present, and Future," p. 45, and Watson and Cook-Deegan, "Origins of the Human Genome Project," p. 10.

22. See the quote in the eprigraph of this chapter.

23. Robert Sinsheimer, "The Santa Cruz Workshop, May 1985," *Genomics* 5 (1989): 954-965.

24. Watson and Cook-Deegan, "Origins of the Human Genome Project," p. 8.

25. Ibid., pp. 8-9.

26. Watson, "The Human Genome Project: Past, Present, and Future," p. 45.

27. Cook-Deegan, "The Human Genome Project: The Formation of Federal Policies in the United States, 1986-1990," p. 112.

28. Renato Dulbecco, "A Turning Point in Cancer Research: Sequencing the Human Genome," *Science* 231 (7 March, 1986): 1055-1056.

29. See United States, Department of Health and Human Services, Public Health Service, National Institutes of Health and United States, Department of Energy, Office of Energy Research, Office of Health and Environmental Research, *Understanding Our Genetic Inheritance; The U.S. Human Genome Project: The First Five Years, FY 1991-1995*, DOE/ER-0452P (Washington, DC: National Institutes of Health, U.S. Department of Health and Human Services and U.S. Department of Energy, April 1990), Appendix 4, pp. 47-50.

30. Cook-Deegan, "The Human Genome Project: The Formation of Federal Policies in the United States, 1986-1990," p. 102.

31. See D. J. Weatherall, *The New Genetics and Clinical Practice*, 3d ed. (Oxford: Oxford University Press, 1991), p. 3, where the "new genetics" is defined simply as "the study of inheritance at the molecular level."

32. See Victor McKusick's discussion of the four methodological streams behind these developments in Victor A. McKusick, "Current Trends in Mapping Human Genes," *The FASEB Journal* 5, no. 1, (January 1991): 14.

33. The first suggestion for creating a complete linkage map came from David Botstein, "Construction of a Genetic Linkage Map in Man Using Restriction Fragment Length Polymorphisms," *American Journal of Human Genetics* 32, (1980): 314-331.

34. Weatherall, *The New Genetics and Clinical Practice*, pp. 120-122.

35. Ibid., pp. 123-125.

36. Ibid., pp. 83-85.

37. Ibid., pp. 91-93.

38. It is not always clear what should count as a "genetic disease" and certain researchers have a stake in defining clear boundaries around therapeutic and non-therapeutic applications. So, perhaps, does the federal government. We will come back to this question below.

39. Weatherall, *The New Genetics and Clinical Practice*, p. 1.

40. C. Thomas Caskey and Victor A. McKusick, "Medical Genetics," *JAMA* 263, no. 19 (May 16, 1990): 2654-2656. The number is now expanding at a rate of almost ten genes per month.

41. For my purposes, a threshold can be defined rather non-technically as that point at which a certain effect begins to be produced.

42. Victor A. McKusick, "Current Trends in Mapping Human Genes," p. 17.

43. Beth Matter, "Mapping the Human Genome: Will It Solve the Mystery of Life?," *Vanderbilt Magazine* (Fall, 1993), p. 18.

44. Quoted in Lee, *The Human Genome Project: Cracking the Code of Life*, p. 150.

45. Note that an "acceptable time-frame" can be defined in various ways. It means "as soon as possible" when clinicians discuss their affected patients, while it means "at any rate faster than European or Japanese researchers" when investigators or politicians enter the conversation.

46. I am not alone in viewing this project as a social experiment. See Eric T. Juengst, "Human Genome Research and the Public Interest: Progress Notes from an American Science Policy Experiment," *American Journal of Human Genetics* 54 (1994): 121-128.

47. I believe the biomedical applications expected of HGP are the primary or most important reasons behind the congressional decision to fund the project, with the economic concerns a close second. But this is certainly an arguable point. Compare, for example, George J. Annas and Sherman Elias, "The Major Social Policy Issues Raised by the Human Genome Project," in *Gene Mapping: Using Law and Ethics as Guides*, eds. George J. Annas and Sherman Elias (New York: Oxford University Press, 1992), p. 5. They believe Congress funded HGP *primarily* for economic reasons.

48. Quoted in Cook-Deegan, "The Human Genome Project: The Formation of Federal Policies in the United States, 1986-1990," p. 119.

49. See, for example, Willard Gaylin, "Fooling with Mother Nature," *Hastings Center Report* (January/February 1990): 17-21; and, George J. Annas, "Mapping the Human Genome and the Meaning of Monster Mythology, *Emory Law Journal* 39, no. 3 (Summer 1990): 629-664.

50. This argument is informed by Charles E. Lindblom, *Politics and Markets: The World's Political-Economic Systems* (New York: Basic Books, Inc., Publishers, 1977).

51. This is perhaps one way my interpretation of HGP's funding may differ from Cook-Deegan's. He interprets Congress primarily as *reacting* to previous research. Undoubtedly, it was reacting to previous research, but it was also reacting to research that it had encouraged with earlier funding. It thus not only reacts to earlier basic research, it helps to *direct* future research. Thus, where Cook-Deegan puts the emphasis on new scientific breakthroughs, I might put the emphasis on the state sponsorship (though both are important).

52. For a historical interpretation of this phenomenon, see Robert Wuthnow, "The Institutionalization of Science," in *Meaning and Moral Order: Explorations in Cultural Analysis* (Berkeley: University of California Press, 1987), pp. 265-298.

53. See Watson and Cook-Deegan, "Origins of the Human Genome Project," p. 10.

54. Rosemarie Tong, *Ethics in Policy Analysis*, (Englewood Cliffs, NJ: Prentice-Hall, Inc., 1986), pp. 14-29.

55. Some of these methods compete with cost-benefit analysis, while others supplement it. For example, there are a host of other economic approaches, such as cost-effectiveness analysis, risk assessment, and risk-benefit analysis. There are also a growing number of non-or quasi-economic approaches, such as technology assessment, regulatory and legal analysis, environmental and political impact assessment, policy evaluation and, as with HGP's ELSI program (discussed below), ethical analyses. See Vincent Vaccaro, "Cost-Benefit Analysis and Public Policy Formation," in *Ethical Issues in Government*, ed. Norman E. Bowie, (Philadelphia: Temple University Press, 1981) pp. 146-162.

56. For background to this claim, see Jurgen Habermas, *The Structural Transformation of the Public Sphere: An Inquiry into a Category of Bourgeois Society*, trans. Thomas Burger, with assistance from Frederick Lawrence (Cambridge, MA: The MIT Press, 1989). For a more current treatment, from the US perspective, see Robert N. Bellah, Richard Madsen, William M. Sullivan, Ann Swidler, and Steven M. Tipton, The Good Society (New York: Alfred A. Knopf, Inc., 1991). I will not address here the question of whether the government ought to be so involved in the economic sphere, though I do not see how it could now withdraw without massive social upheaval.

57. The distinction between public and private is a controversial one, especially for some feminists. I intend by it simply an analytical distinction between sources of funding.

58. This summary is dependent on Peter H. Aranson, *The Political Economy of Science and Technology Policy*.

59. Vaccaro, "Cost-Benefit Analysis and Public Policy Formation," p. 150.

60. Henry Rowen, "The Role of Cost-Benefit Analysis in Policy Making," in *Cost Benefit Analysis and Water Polution Policy*, eds. Henry M. Peskin and Eugene P. Seskin, (Washington, DC: Urban Institute, 1975), p. 363. Quoted in Vaccaro, "Cost-Benefit Analysis and Public Policy Formation," p. 146.

61. For a feminist critique of end-state analyses, see Young, *Justice and the Politics of Difference* (Princeton, NJ: Princeton University Press, 1990), pp. 15-38. I am persuaded that her critique is a helpful corrective to the typical use of these methodologies, but not that it undercuts their use entirely; rather, this critique and others like it help to define the proper limits of these economic tools.

62. On the other hand, a truly educated guess is better than a mere guess.

63. Aranson, *The Political Economy of Science and Technology Policy*, p. 10.

64. National Research Council, *Mapping and Sequencing the Human Genome*, pp. 90-91.

65. United States, Congress, Office of Technology Assessment, *Mapping Our Genes: Genome Projects: How Big, How Fast?*, p. 185.

66. National Institutes of Health and United States Department of Energy, *Understanding Our Genetic Inheritance: The U.S. Human Genome Project: The First Five Years, FY 1991-1995*, p. 16.

67. Cook-Deegan, "The Human Genome Project: The Formation of Federal Policies in the United States, 1986-1990," p. 124. Note that this is not the same one thousand year figure used by Watson. This figure describes the projected length of time needed to *produce* a complete set of genome maps at a rate of two million base-pairs per year, while Watson's figure describes the projected time required to *interpret* these maps *after* HGP (in its present form) has completed its fifteen-year mission.

3

Diagnostic and Therapeutic Applications of Human Genome Research

> [T]he major commitment...we need to make in starting this project is to see it through. Because it is in the final stages that we will...see some of the fruits of this labor. If we end in this halfway zone, then I think we are in deep trouble...[I]t...could be a potentially disastrous situation, to sit in hiatus between detection and cure...[1]
>
> Nancy S. Wexler

> [T]he human genome initiative will...invite attack from those who are fearful or hostile toward the future. It should also attract the active support of those willing to defend the future.[2]
>
> David Baltimore

Money and time are not the only costs that need to be estimated when undertaking a project with the scope and reach of the US Human Genome Project. There are costs of a different sort that individuals and society may have to bear as a result of this project's existence, and the maximizing of efficiency in order to save time and money may in fact make these costs to individuals and society that much greater. At least, this is what I will suggest in this chapter; moreover, this chapter argues that many of these other costs will be born by future generations, and that they will be born unequally between generations. That is, informed observers expect that HGP's future applications will be developed in three overlapping, multi-generational phases, with the majority of HGP's future costs expected to occur in the second or "interim" phase. This troubling future is directly related to the nature of the research itself and to the decision to make HGP a "big science" or "dedicated" project.

We learned in the previous chapter that HGP is commonly and correctly viewed as a basic research project, but that it was funded by Congress principally for the applied technologies it is expected to spin off or make possible, especially those biomedical technologies that will lead to the detection, prevention, and treatment of literally thousands of genetic diseases and disorders, about which little can currently be done. Moreover, with thousands of genetically-based tests and therapies expected to emerge from HGP's research, Congress was not slow to realize its possible economic and political implications, and these implications played a key role in the decision to fund the project as well.

We also learned that in justifying HGP Congress made some rather vague appeals to, or on behalf of, future generations. These appeals need not imply that currently living persons will not also benefit from HGP's research. Some researchers and private biotechnical firms will enjoy certain types of near-term benefits as a result of gaining access to HGP's funding, and a limited number of patients in this generation will enjoy the benefits (and endure the burdens) of HGP's early applications. Nevertheless, future generations are expected to enjoy the vast *majority* of HGP's applications.

This chapter will investigate some of the social and ethical implications of this claim. Specifically, I will inquire whether there are grounds for Congress' belief that future generations will benefit from this research. I also want to explore in more depth how the two applications of human genome research mentioned in chapter 1, preimplantation diagnosis of genetic disease and germ-line gene therapy, will give agents unprecedented control over the very existence of some future persons. These concerns require us to engage in some predictive exercises, however limited the results of these exercises may be.[3] It was just such exercises that led HGP's advocates to decide to constrain their own research.

HGP'S COLLECTIVE IMPLICATIONS

Interestingly, and perhaps not surprisingly, the congressional debate concerning HGP did not end with the decision to fund the project. In fact, it extended over several years, and the review in the previous chapter of the arguments behind the goals and shape of HGP summarizes merely the early part of the debate. This early debate took place primarily between scientific and policy elites who had access to the relevant technical information and were able to make use of it. One of the more interesting aspects of this stage of the debate, as noted above, was the concentration on the question of means. Its goals or ends, that is, the new knowledge HGP will produce and the applications it will spin-off, were never really at issue as far as the congressional justification of the project was concerned. Certainly, the question of ends was raised in the sense that HGP's goals and benefits were enumerated and categorized, but it was not raised in the sense that I intend here, namely, that these ends were justified by some sort of explicit argument.[4] Rather, the early debate centered largely on the question of which means would be the most time- and cost-efficient in the long-term.

However, as news of the nascent project filtered into the popular and scientific press, the debate behind HGP began to include more diverse types of participants with more diverse types of concerns. While it must acknowledged at the outset that HGP went forward in large part as planned, the broadening of the debate behind it eventually required advocates of the project to consider, with Congress, a growing public concern for HGP's long-range "social costs."[5]

THE "ELSI HYPOTHESIS"

Among HGP's early advocates, it was James Watson himself who surprised his colleagues by unexpectedly proposing that HGP researchers acknowledge and address public concerns relative to the project. Watson had already played a key role in the technical debate behind HGP and was subsequently named as the first head of the National Center for Human Genome Research (NCHGR) at the NIH. He backed his proposal by announcing at a congressional hearing[6] that 3% or even more of NCHGR's research budget would be directed toward the study the project's ethical, legal, and social implications.[7] Watson later reported that his unprecedented proposal was intended to head off a "strong popular backlash" toward the genetic research community.[8] Whatever his personal motivations, however, Congress quickly warmed to his suggestion and, at the urging of then Senator Albert Gore and Senator John Kerry, insisted that the more reluctant officials at DOE fund a s imilar program.[9]

This suggestion, coming as it did from a respected scientific source and backed by strong congressional support, led to the formal institution under NCHGR of the Ethical, Legal, and Social Implications (ELSI) Program. Its unfortunate acronym notwithstanding, the so-called "ELSI hypothesis" embodies the lofty conviction that

> ...combining scientific research funding with adequate support for complementary research in the social sciences and humanities will help our social policies about science evolve in a well informed and authoritative way.[10]

Specifically, ELSI has a fourfold mission: 1) to anticipate and define the implications expected of HGP for individuals and society; 2) to examine the ethical, legal, and social sequelae of mapping and sequencing the human genome; 3) to stimulate public discussion of the issues thus identified; and 4) to develop policy options to assure that the information produced by HGP is used for the benefit of the individual and society.[11] In short, ELSI's mission is to anticipate problems that could arise as a result of mapping and sequencing the human genome and to suggest educational and policy options that are designed to forestall or minimize the adverse effects of these problems.[12]

Of course, whether ELSI researchers will realize their ambitious goals remains to be seen. Here, I will briefly review the specific concerns that ELSI researchers and other informed observers have identified thus far, and then concentrate on the expected "sequelae" associated with human genome research in order to identify what I take to be the key issues emerging from this project with regard to future generations.

ETHICAL AND SOCIAL IMPLICATIONS OF HGP

Two early and previously mentioned studies on the proposed project, one by the NRC and the other by the OTA, include brief comments on the ethical, legal, and social implications expected of HGP,[13] but the first comprehensive list of concerns was drawn up by the Working Group on the Ethical, Legal, and Social Issues Related to Mapping and Sequencing the Human Genome. It is published as "Appendix 7" in the NIH's and the DOE's *First Five Year Plan* for HGP.[14] In this report, ELSI concerns are listed under nine "topics" deemed by the Working Group to be of "particular importance." These nine topics include the following concerns.

> Fairness in the use of genetic information with respect to: insurance (acquisition and maintenance of health, life, disability); employment (equal access); the criminal justice system; the educational system; adoptions; the military; [and] other areas to be identified.
>
> The impact of knowledge of genetic variation on the individual, including issues of: stigmatization; ostracism; labelling; [and] individual psychological responses, including impact on self image.
>
> Privacy and confidentiality of genetic information regarding: ownership and control of genetic information; [and] consent issues.
>
> The impact of the Human Genome Initiative on genetic counseling in the following areas: prenatal testing; presymptomatic testing; carrier status testing, especially for very common disorders such as cystic fibrosis; testing when there is no therapeutic remedy available, such as for Huntington's disease; counseling and testing for polygenic disorders; [and] population screening versus testing.
>
> Reproductive decisions influenced by genetic information: effect of genetic information on options available; [and] use of genetic information in the decision-making process.
>
> Issues raised by the introduction of genetics into mainstream medical practice: qualifications and continuing education of all appropriate medical and allied health personnel; standards and quality control; education of patients; [and] education of the general public.

> Uses and misuses of genetics in the past and the relevance to the current situation, e.g.: the eugenics movement in the U.S. and abroad; problems arising from screening for sickle-cell trait and other recent examples in which screening or testing sometimes achieved unintended and unwanted outcomes; [and] the misuse of behavioral genetics to advance eugenics or prejudicial stereotypes.
>
> Questions raised by the commercialization of the products from the Human Genome Initiative in the following areas: intellectual property rights (patents, copyrights, and trade secrets); property rights; impact on scientific collaboration and candor; [and] accessibility of data and materials.
>
> Conceptual and philosophical implications of the Human Genome Initiative on: the concept of human responsibility; the issue of free will versus determinism; [and] the concept of genetic disease, particularly in view of the high rate of human genetic variability and the large numbers of people who will be found to have genetic vulnerabilities.[15]

Subsequently, the implications for health care were delegated to the NIH's ELSI program and those that concern science education and commercialization were delegated to the DOE's Office of Health and Environmental Research.[16]

While the above list defies easy summarization or categorization,[17] this much has been observed. First, none of the above issues can be said to represent a genuinely "new" problem for our society.[18] That is, ELSI's Working Group believes we have encountered each of these issues in the past as a result of previous genetic research efforts. Second, each issue is oriented in some way toward the possible use or misuse[19] of the genetic information stemming directly from HGP's expected *applications*, and thus only indirectly from the project as such.[20] Certainly the potential for the misuse of genetic information is not new. One need only point to the abuses of the Nazis before and during World War II or, in this country, to early sickle cell screening efforts that went awry. No one believes that these types of abuse are not still possible with the genetic information stemming from HGP. Nevertheless, it may be safe to generalize that

ELSI's concerns have been given a somewhat different focus. That is, ELSI researchers seem to be guardedly optimistic that past abuses can be avoided or minimized with education and careful policy development, and thus they have never advocated an absolute ban or a moratorium on this research. But they do worry that the potential for these abuses will continue and, in fact, will increase dramatically because of the sheer *volume* of information expected from HGP's biomedical applications. Said differently, the basic problem stemming from HGP is thought not to be one of deliberate or intentional abuse by evil or misguided political or medical elites, but a *systemic problem* that arises from *information overload* and is in turn directly related to the massive scale of the project's undertaking. It is *this* concern that lies behind most of the ethical, legal, and social problems expected of HGP and its applications.

A Three-Phase Future

Recall that the massive scale of HGP is thought to be a rational response to what I characterized in the previous chapter with the threshold metaphor; that is, HGP was thought to be the most time- and thus cost-efficient way to bring the biomedical applications envisioned by researchers in the mid-1980s into routine or standard medical use. Even before HGP was started, however, the information output of traditionally-based genetic research was threatening to overwhelm health care systems in the developed countries of the world. HGP is expected to accelerate this information output exponentially.

To get an intuitive sense for this concern, consider again that the ambitious goal of the project's basic research will be successfully accomplished when a *single composite* individual's human genome has been mapped and sequenced. This goal alone is presenting researchers with enormous technical ("informatic") challenges for the secure electronic storage, manipulation, and retrieval of this information.[21] However, as HGP's applications come into routine medical use, it will not be a single human genome that will need to be securely processed, but the test results and the relevant portions of the genomes for thousands of individual *patients*, the information for whom will reside in a medical record. At the upper limit, it is estimated that HGP's biomedical applications will permit as much as *ten million* times more information on the molecular level *per individual* to be accumulated in a given medical record than is now available.[22] It is this information that has observers especially concerned, and it is this information that requires ELSI researchers to look beyond the completion of the fifteen-year project itself to the long-term implications of its expected applications.

Nancy S. Wexler, chair of the Working Group for the ELSI program, anticipates that the project will lead to a future scenario that falls into three broadly defined, overlapping phases. In testimony before the Senate Commerce, Science, and Transportation Committee, she stated:

> The long-range goal of the Human Genome Initiative is treatment, prevention, and cure. But the initial benefits of the science are going to be in detection. And, for a while we are going to be in an interim phase in which you can detect, but you cannot [treat or cure those disorders that are detected].[23]

This scenario is also predicted by the ELSI Working Group itself. They wrote in their report that:

> Although initially the Human Genome Initiative will produce information that will lead to the detection and diagnosis of genetic disease, the long-range goal will go beyond this to providing improved treatment, prevention and ultimately cure. The interim phase, before adequate treatment is available, is the one in which the most deleterious consequences can occur, such as discrimination against gene carriers, loss of employment or insurance, stigmatization, untoward psychological reactions and attention. Once effective treatment is available for an illness, most of these problems disappear.[24]

If we assume a relatively steady level of funding for genome research and its spin-off applications, these statements suggest the following future.

The first phase is a research and development phase that begins (or began) when HGP spins off its first application. This phase is expected to lead rather quickly to a large number of genetic screens and tests that will permit researchers to detect and diagnose directly[25] increasing numbers of known genetic and multifactorial diseases. The second phase is the one that most concerns informed observers; they call it an "interim" phase, and we can only hope they are correct. In this phase, the thousands of new diagnostic screens and tests researched and developed in the first phase will move into the medical mainstream in such numbers as to be counted as "routine" or "standard" care. The notion of routine or standard medical care is key in distinguishing the first and second phases. It, too, will involve some sort of threshold, or at least the perception of one, where these new tests are widely accepted by the public and where physicians can be held liable for not using them. There are a limited number of such tests available and in use today, but most of these are still generally regarded as experimental. Many more, however, are in the research "pipeline" and will soon be ready for clinical trials and use. It is the increasing use of these tests, with increasingly large numbers of patients for increasingly large numbers of diseases, that will lead to the accumulation of massive amounts of potentially "dangerous" diagnostic information.[26] Phase three represents that phase where clinicians will not only be able to detect genetic disease, but also be able to treat it and perhaps even cure it. The ELSI Working Group believes these capabilities will close the therapeutic gap in clinical genetics; further, it believes, perhaps somewhat naively, that closing the therapeutic gap will cause the problems expected in phase two to "disappear."

In response to this projected future, ELSI researchers more recently have focused their efforts on planning for the problematic second phase. Its foreseen problems have been catalogued as types of "choices" for "public and professional deliberation." They concern:

> Choices for individuals and families about whether to participate in testing, with whom to share the results, and how to act on them;
>
> Choices for health professionals about when to recommend testing, how to ensure its quality, how to interpret the results, and to whom to disclose information;
>
> Choices for employers, insurers, the courts, and other social institutions about the relative value of genetic information to the kinds of decisions they must make about individuals;

Choices for governments about how to regulate the production and use of genetic tests and the information they provide, and how to provide access to testing and counseling services;

Choices for society about how to improve the understanding of science and its social implications at every level and increase the participation of the public in science policymaking.[27]

This list may leave the reader with the impression that these problems are discrete, but they are discrete only in analytical terms. In fact, they will cut across societal institutions that are centrally or even tangentially concerned with health care and will affect individuals in ways that are complexly and intricately linked to these institutions. For instance, because of the vast amounts of information generated by these new tests, there is concern that it will be difficult to maintain any semblance of privacy and confidentiality in the medical records of individual patients; or, conversely, that if special efforts are taken to maintain privacy and confidentiality, these efforts will significantly increase the information costs related to health care administration. There will be additional costs associated with training health care providers in the use and interpretation of these tests, and with educating providers, patients, and the public in interpreting the significance of their results, most of which will be given in probabilistic terms.

Moreover, the increasing costs that result from these tests will only compound the problems of access that currently plague our health care system. It is not at all clear who will pay for these tests as they enter mainstream medicine, but there is every reason to believe that the public will perceive them as desirable. In turn, this perception will have complex implications for the demand side of health care. In the effort to keep health-related costs down, employers or health insurance companies may be tempted to use this information to discriminate against employees or clients. Or, again, if employers or insurance companies are mandated by law to cover all individuals without regard to genetic diagnoses, it is likely that their costs will simply be shifted to consumers or tax-payers with uncertain results for overall US productivity and competitiveness. Also, unaffected individuals may learn that they do not need certain types of health insurance coverage, and this could create a smaller risk pool. The overall effect of these problems may be to create a class of genetically vulnerable people who could easily face discrimination, stigmatization, and psychological trauma on the basis of their genetic make-up.[28]

A STATISTICALLY-BASED THERAPEUTIC GAP

ELSI researchers are not, however, simply concerned with the number of new genetic diagnostic tests and the concomitant volume of information stemming from them. They are also worried about what these tests substantively can do or not do. What they can do is detect and help diagnose the presence of, or probability of developing, genetic disease. What they cannot do, with certain qualifications to be supplied momentarily, is treat or cure these diseases. Thus, during the second phase of this probable future, these tests will greatly exacerbate the long-standing "therapeutic gap" in clinical genetics. This has also been judged to be one of the more significant concerns stemming from HGP research. Wexler, for example, is very concerned that politicians will begin this project and then, when they see second phase problems mounting, withdraw their support for funding.[29]

The meaning of the term *therapeutic gap* in genetics is intuitively obvious when it is used in reference to a single disease, one that can be detected and diagnosed but not adequately treated or cured. Its use in this context, however, requires additional qualification. I am using the term to describe not just the gap between the ability to diagnose and to treat a single disease, but as a statistical product representing the *cumulative effect* of thousands of such gaps. Is it possible that this cumulative effect would push us toward a threshold of problems judged to be genuinely new?

I have suggested elsewhere[30] that the claim that HGP will lead to no new problems can imply a number of things. It can imply, as we saw above, that our society has already encountered the problems ELSI is designed to investigate due to previous genetic research efforts, and this is undoubtedly true. However, because these problems are not new, this claim may also suggest that we will not therefore need new solutions to them, that is, that solutions to these problems will not require new ethical, legal, or social *paradigms*, but merely the educational efforts and policy adjustments envisioned by ELSI researchers. Construed in this second way, it does not follow that HGP's research will occasion no new problems for our society. Because we do not expect to see new problems does not mean that "old" problems, when they emerge under new circumstances, will not require new solutions.[31]

Another qualification of the term *therapeutic gap* is also required. The meaning of "genetic disease" is ambiguous, and it is applied to a range of disorders. Clinically, genetic disease tends to be grouped under several major and largely functional categories, which include single gene or monogenic disorders, chromosomal disorders, congenital malformations or other common diseases such as cancer in which genetic predisposition plays a large part (multi-factorial diseases), disorders of the mitochondria, and disorders due to random somatic cell mutations.[32] A functional approach to genetic disease is particularly problematic, however, for those disorders in which the patient shows no—symptoms—those categories of genetic disease for which the patient is a carrier, is predisposed, or is merely presymptomatic. Moreover, disease is a socially constructed category, and the presence or absence of certain human capacities, based in genetic functioning, is not enough to define the term adequately.

As a result of this definitional ambiguity, the line between what counts as diagnosis and treatment of genetic disease is not always clear. In short, the effects of diagnosis are not uniform across all genetic diseases, being more helpful for some and more problematic for others. Thus, for certain genetic problems, such as having a predisposition to certain types of colon cancer, detection alone may be genuinely helpful. Testing regimens can be undertaken to detect the early symptoms of disease, and lifestyle changes can be implemented that may forestall these symptoms or reduce their severity.[33] However, for other types of genetic disease that are now or soon will be detectable presymptomatically, the classic example being late-onset Huntington's disease, such detection may be more of a burden than a benefit for individuals and their families. There is no known treatment for this or most genetic diseases and, as mentioned above, those few treatments that do exist are of questionable effectiveness and very costly.

Carriers of genetic disease raise similar concerns. Whether knowledge of one's carrier status is beneficial or burdensome is a highly individualistic determination, one that is influenced by the type and severity of genetic disease in question, whether the carrier's spouse is a carrier of the same disease and the couple is planning to have children, and, for sex-linked diseases, whether a fetus that is conceived by parents who are carriers for such diseases is male or female. It is true that these are not new problems. But they are expected to be radically "democratized" in the second phase of this probable future and spread across a considerably wider population.

LINGERING EMPIRICAL UNCERTAINTIES

Do we know when this troublesome second phase will begin or how long it will last? Eric Juengst, then acting director of the ELSI program at the NIH, says only that:

> The most immediate consequence of genome research will be the development of new diagnostic and predictive tests, *well in advance* of corresponding therapeutic or curative advances.[34]

He does not define "well in advance," though we learned above that the second phase will begin when enough tests are in mainstream medical use to count as routine or standard medical care. Thus, it is probably safe to generalize that the second phase will begin at some point after the conclusion of the fifteen-year basic research project, when the human genome is largely mapped and researchers are moving into the arduous task of interpreting and confirming the function of the genetic code. Some researchers are, of course, already seeking the location of and developing tests and therapies for certain "high-profile" disease genes; thus particular research and development breakthroughs may occur at any time. Nevertheless, as a cumulative or statistically-based product for our society, this second phase remains a future scenario, albeit a scenario whose onset may be in the relatively near future.

How long will the second phase last? This question is even more difficult to answer, but we may surmise that it will last as long as researchers are learning new knowledge about the structure of the genome and its functions, and are spinning off new applications on the basis of this information. Fortunately, over time the cumulative negative effects of these new discoveries should begin to diminish. Again, however, if James Watson's enthusiastic prediction that researchers will need an additional one thousand years to unravel the mysteries of the human genome is accurate only by a quarter or even a tenth, our society could be facing a one hundred to two hundred fifty year future of such problems.[35]

Lastly, we can question whether the therapeutic developments expected in the third phase will actually cause the second phase problems to "disappear," as the ELSI Working Group seems to assume. These therapies, at least initially, will be ameliorative rather than curative, and collectively this may work simply to increase the frequency of genetic disease in future generations of patients. Also, those treatments that are in fact developed will also need to be cost-effective enough to permit wide access to them. Early returns with experimental genetic therapies have been encouraging in medical terms, but they are very costly and still far from anything that might be called routine medicine.

IMPLICATIONS FOR INDIVIDUAL CHOICES

We have been considering the collective implications of HGP and its expected applications. It must be remembered, however, that these collective social costs will be driven not only by research developments but also by thousands of individual patients who will or will not use HGP's genetic applications. Moreover, in many cases it will be these same individual patients who will ultimately bear the costs and burdens of these collective implications. Thus, we must pause here and introduce at least a few examples of the kinds of problems and types of choices individuals will face with respect to future generations as a result of HGP having been implemented as public policy.

I will use the preimplantation diagnosis of genetic disease to illustrate the kind of *diagnostic* capabilities that are now in clinical use experimentally and expected to be in routine use in phase two; and I will use germ-line gene therapy, which is currently not feasible in humans but which may be in the near future, to illustrate a possible *therapeutic* application that could be widely available in phase three. One could, of course, choose other applications to illustrate these issues, but the two I have chosen will be useful for present purposes because of the peculiar implications they hold for evaluating HGP's effects on future generations.[36]

BACKGROUND: IVF TECHNOLOGIES

It is likely that HGP's research will result in literally thousands of new clinical applications. These applications will be used in a multitude of ways, each with its own special problems and possibilities. However, many, if not most, of HGP's early applications are likely to be utilized in very early stages of the human reproductive cycle. These applications will thus be possible not only because of the basic research emerging from HGP, but also because of the techniques and procedures that have emerged from earlier and ongoing *in vitro* fertilization (IVF) research.[37]

In vitro fertilization— literally "fertilization in glass"—was tried with animals as early as the late nineteenth century, and was widely used in animal breeding by the 1960s. IVF was developed and first used with humans by 1965 to research and treat various types of infertility, and its use resulted in the first live human birth in 1978 (the first "test-tube" baby). By 1990, approximately 38,000 babies had been born world-wide using IVF.[38] Also, its use has dramatically expanded in the attempt to research new approaches to contraception and, of concern here, to the prenatal detection of genetic and congenital abnormalities.[39]

Technically, IVF is merely one step in a multi-step procedure, which begins with the stimulation of the ovaries and the recovery of multiple eggs or oocytes (multiple eggs are recovered in order to increase pregnancy rates or to have extra eggs on hand for research purposes). Sperm is added to the eggs *in vitro* and, under the proper conditions, fertilization takes place in about twenty-four hours (there is no "moment" of conception). Fertilization under these conditions is successful about 80% of the time.[40] After fertilization, the preimplanted embryo, or "preembryo," is permitted to mature for another forty-eight to seventy-two hours, at which point it is transferred to a woman's uterus (again, several are usually transferred if they are available in order to increase pregnancy rates). At the point of transfer, the preembryo is usually between two and eight cells.[41] After transfer there is roughly a 30% chance per transfer and a 10% chance per embryo that pregnancy will occur,[42] though these percentages may increase with experience.

Within the multi-step IVF procedure just described, geneticists have concentrated on that period after *in vitro* fertilization and before transfer, when the *in vitro* preembryo can be cultured, biopsied, and cryopreserved both for research and for diagnostic purposes.[43] It is also at this point that therapeutic interventions may be attempted in the future for those couples who do not wish to discard those preembryos found to be genetically defective in some way.

PREIMPLANTATION DIAGNOSIS

The preimplantation diagnosis of genetic disease relies on the micromanipulation techniques developed in IVF research and utilizes a variety of techniques developed in the last decade by genetic researchers, such as polar body biopsy, multicell biopsy, blastocyst biopsy, karyotyping of chromosomes, *in situ* hybridization, and polymerase chain reaction. Clinicians also utilize techniques that can detect alterations in gene structure and, to a limited extent, in gene products.[44] The procedure also permits clinicians to sex the preembryo and, for various sex-linked genetic diseases, to transfer or discard the relevant preembryos on this basis. HGP is expected especially to expand those techniques that detect alterations in gene structure at the molecular level. Generally, these techniques involve extracting genetic material from the preembryo (different tests are employed at different points of development) and testing it for the presence of genetic disease. Within certain technical limits, this procedure has no adverse effects for the future child born as a result of the procedure.

Thus viewed, preimplantation diagnosis is seen by most observers as an extension of IVF technology,[45] though it is aimed not at infertile couples but at fertile couples who know they are at risk of passing serious disease genes to their offspring. Unfortunately, many of these couples have already given birth to one or more affected children, which is usually how they learn they are at risk. The advantage of preimplantation diagnosis for a couple in this situation is that it greatly decreases the probability that they will give birth to another affected child (though this outcome is not guaranteed, especially if geneticists are testing for a particular disorder). It also avoids the maternal risks and emotional trauma of an abortion should the fetus be tested prenatally by chorionic villus sampling (at eight weeks) or amniocentesis (at sixteen weeks), and determined to be affected.[46]

It is not necessary to explore here the technical advantages and limitations of each of the above-mentioned tests.[47] Likewise, their use raises a number of ethical issues with respect to safety, efficacy, cost, and access that are of genuine but not dissimilar concern to other experimental biotechnologies. Here I want to note what I think is the unique, if obvious, feature of preimplantation diagnosis with respect to future generations: it offers the relevant agents (couples, researchers, and clinicians) the opportunity to choose between those preembryos to be implanted and those to be frozen, discarded, or used for research and then discarded. It thus greatly expands the power and degree of control that current generations can exercise over future individuals. Within certain technically-limited probabilities, it gives agents the power to choose who will live and who will not live. This expansion of power and control raises what, in my estimation, is the key ethical question of this application for justifying actions that affect future generations: on what *moral* basis will such a choice be made and justified? Germ-line genetic therapy raises the same question in a somewhat different context.

Germ-line Gene Therapy

In addition to the diagnostic applications expected of HGP, geneticists are also seeking to develop effective therapeutic applications. However, we learned above that many of these therapeutic applications are likely to be realized much further in the future than many of the diagnostic applications. This is one source of the perennial therapeutic gap in clinical genetics. There is some obvious benefit to detecting and diagnosing those genetic diseases where changes in lifestyle or diet can significantly affect the patient's prognosis (for example, phenylketonuria or PKU), but many diagnostic options will be of little or no help to the approximately 1% of all children born who suffer from genetic problems for which treatments are non-existent or are severely limited in their effectiveness. Human genome research offers hope that this situation can be changed, in part, by helping researchers understand disease processes better on the molecular level and, in part, by helping them develop therapies that are not now available. Some of these therapies could involve the manipulation of genetic code in the preembryo, resulting in changes that are passed both to the future child that is born as a result of the procedure or intervention and to future descendants of this child.

Gene therapy can be generally defined as a

> medical intervention based on modification of the genetic material of living cells. Cells may be modified *ex vivo* for subsequent administration to humans, or may be altered *in vivo* by gene therapy given directly to the subject.[48]

One of the leading advocates and researchers of gene therapy is W. French Anderson. He has proposed a basic categorization of genetic therapies that distinguishes between somatic cell gene therapy and germ-line gene therapy.[49]

Somatic cell gene therapy refers to the insertion of a normal gene or DNA fragment into somatic or body cells of the patient. Ideally, only the patient is affected by this therapy, and changes produced by the therapy would not normally be passed to the patient's descendants.[50] This implies that if researchers developed a treatment for a disease that permitted the patient to live and reproduce, the treatment would need to be repeated for each succeeding generation.[51] The first NIH-supported somatic cell gene therapy began in September 1990 with a number of young adenosine deaminase deficiency (ADA) patients.[52] There are ongoing concerns about this therapy, due primarily to its experimental status, but it is generally agreed that it does not differ in kind from other types of therapeutic interventions. Somatic cell therapy differs only in the technology being utilized and in the sites of the intended intervention. Also, it seems to be broadly supported by public opinion. A recent poll reported by the OTA indicates that 83% of the public supports somatic cell therapy to cure usually fatal diseases.[53]

As its name implies, germ-line gene therapy is aimed specifically at the reproductive or germ cells, though it will utilize many of the same technologies that somatic cell gene therapy utilizes. At this site, it will correct or replace the defective gene(s) or DNA fragment with foreign or engineered genetic material. Thus, germ-line gene therapy can be distinguished from somatic cell gene therapy not by the types of actions performed, but by the intended ends of the actions. It will not only ameliorate symptoms for the given "patient" (which, in fact, will usually be a preembryo), but also remove the physical *cause* of the problem for that patient and for all future descendants of the patient (insofar as a given gene or DNA fragment can be said to "cause" a disease). Germ-line gene therapy thus may also be distinguished by its intended or direct effects on future generations.

Germ-line gene therapy is not technically feasible in humans and the NIH does not currently consider research protocols that involve germ-line manipulations in humans.[54] Moreover, some observers have suggested that preimplantation diagnosis will make the much more technically difficult germ-line gene therapy unnecessary. Others, however, believe that it is simply a matter of time before researchers will want to try it in humans. The ability to eliminate a truly and unquestionably lethal disease gene (for example, the gene responsible for Tay-Sachs disease) will continue to present a compelling reason for some to move ahead with this research.[55] Nevertheless, it remains true that germ-line gene therapy does not currently enjoy the same widespread acceptance that somatic cell gene therapy does, and its application to humans continues to be vigorously debated in the literature. While it too raises ethical issues similar to other new genetic and non-genetic biotechnologies (safety, efficacy, cost, and access), most of the objections to it arise from its unpredictable consequences for future generations. Thus, perhaps better than any other application of HGP, it illustrates the need to grapple with the increasing powers this research will give us. Again, this therapy raises the question: on what *moral* basis will we choose between the various options germ-line therapy makes available to us? Which, if any, genetic problems will we try to fix permanently for future persons; that is, which problems will we try to eliminate from the gene pool and thus, which persons will live and which will not? How will we decide such questions now that we are confronted with them, as perhaps we have never been before in the history of humankind? We take up these questions in the next chapter. Before before we do, however, I want to summarize our findings thus far.

Preliminary Findings

Chapters 2 and 3 have addressed the first basic aim of the book, namely, to investigate *how* HGP will affect future generations. I believe we have identified at least two ways HGP will affect future generations. The first originates on the collective level and results from the likely three-phase inter-generational allocation pattern, which in turn is directly related to the "dedicated" or "big science" shape of the project. If we can assume that a steady stream of funding will be maintained, and that researchers' predictions are at least roughly accurate, HGP's basic science goal will be reached within the next fifteen years. We learned, however, that it may take a century, and perhaps several centuries, of further research to interpret this data and to develop enough therapeutic applications (phase three) to offset the social costs produced by the research and development of HGP's diagnostic applications (phase two). Thus, because of the cumulative effects of this "therapeutic gap" and because of the likely "sequelae" of development for HGP's clinical applications—that is, because we are likely to see diagnostic applications developed "well in advance" of the corresponding therapeutic applications—future generations will not benefit from HGP equally. The three-phase future projected by ELSI researchers suggests that future generations living during phase two may bear a much higher share of HGP's social costs than those future generations living during or after phase three. The ethical question this future scenario raises in one of distributive justice; that is, we need to ask whether and on what grounds this probable allocation of benefits, costs, and risks is justified.

Second, on the individual level, the existence, numbers, and identities of some future persons will be contingent on choices made in the context of some of the applications enhanced by or derived from HGP's research. I illustrated these choices in chapter 1 with two cases from clinical genetics, and we explored more fully in this chapter the background technologies that either make or will make them possible. However, because these applications involve HGP's applications in the problem of contingent future persons, it is not clear on what moral basis these choices should be made, especially if we care about future persons enough to try to protect them from predictable genetic harm. For in the context of the two applications reviewed here, the preimplantation diagnosis of genetic disease and germ-line gene therapy, a decision to protect a *particular* future person from predictable genetic harm will mean that a *different* person will in fact be born. In other words, the decision to protect a particular future person from being born with a diagnosed genetic defect will mean either not implanting the preembryo that will lead to that future person or implanting it only after germ-line genetic therapy has altered its future identity. Thus, as will see more fully in the next chapter, our moral reference cannot be the original future person, for as a result of our choice that person will never exist. Again, this is the problem of contingent future persons as it arises in the context of two of HGP's applications. We must now turn, then, to the problem of contingent future persons and to the second basic aim of the book: the *evaluation* of HGP's likely effects on future generations.

NOTES

1. United States, Congress, Senate, Committee on Commerce, Science and Transportation, Subcommittee on Science, Technology and Space, *Hearing on the Human Genome Initiative and the Future of Biotechnology*, S. Hrg. 101-528 (Washington, DC: U.S. Government Printing Office, 1989), p. 98.

2. Quoted in Thomas F. Lee, *The Human Genome Project: Cracking the Genetic Code of Life* (New York: Plenum Press, 1991), p. 231.

3. This predictive exercise raises the questions whether such predictions are in principle possible and, if so, with what degree of confidence. These questions have not, to the best of my knowledge, been discussed in much depth relative to HGP. Two sources that briefly address the issues are Ray Moseley, Lee Crandall, Marvin Dewar, David Nye, and Harry Ostrer, "Ethical Implications of a Complete Human Gene Map for Insurance," *Business and Professional Ethics Journal* 10, no. 4 (Winter 1991): 69-82; and Dorothy Nelkin, "Genetics and Social Policy," *Bulletin of the New York Academy of Medicine* 68, no. 1 (January-February, 1992): 135-143.

4. The exception to this claim was the argument by NIH-sponsored researchers that the DOE's interest in tracking human mutation was not a sufficient reason to undertake a large scale genome project.

5. By social costs, I refer to: 1) the indirect institutional costs that result from this research for health care providers, private and public employers, the health insurance industry, and individuals generally; and, 2) the risks it poses for certain kinds of harm to individuals or groups of individuals. These social costs will be detailed below.

6. United States, Congress, Senate, Committee on Commerce, Science and Transportation, Subcommittee on Science, Technology and Space, *Hearing on the Human Genome Initiative and the Future of Biotechnology*, S. Hrg. 101-528, p. 18.

7. Funding is currently up to approximately 5% of the NCHGR's research budget. See Annas and Elias, "The Major Social Policy Issues Raised by the Human Genome Project," in *Gene Mapping: Using Law and Ethics as Guides*, p. 5. Congress is also considering the creation of a standing commission modeled along these lines. See United States, Congress, House, Committee on Government Operations, *Designing Genetic Information Policy: The Need for an Independent Policy Review of the Ethical, Legal, and Social Implications of the Human Genome Project*, House Report 102-478 (Washington, DC: U.S.

Government Printing Office, 2 April 1992); and, United States, Congress, Office of Technology Assessment, *Biomedical Ethics in U.S. Public Policy—Background Paper*, OTA-BP-BBS-105 (Washington, DC: U.S. Government Printing Office, 1993).

8. Watson, "The Human Genome Project: Past, Present, and Future," p. 46.

9. United States, Congress, Senate, Committee on Commerce, Science and Transportation, Subcommittee on Science, Technology and Space, *Hearing on the Human Genome Initiative and the Future of Biotechnology*, S. Hrg. 101-528, p. 48.

10. Watson and Juengst, "Doing Science in the Real World: The Role of Ethics, Law, and the Social Sciences in the Human Genome Project," p. xix.

11. National Institutes of Health and US Department of Energy, *Understanding Our Genetic Inheritance: The U.S. Human Genome Project: The First Five Years, FY 1991-1995*, p. 21.

12. Ibid., p. 67.

13. National Research Council, *Mapping and Sequencing the Human Genome*, pp. 100-103, and Office of Technology Assessment, *Mapping Our Genes: Genome Projects: How Big, How Fast?*, pp. 79-88.

14. National Institutes of Health and US Department of Energy, *Understanding Our Genetic Inheritance: The U.S. Human Genome Project: The First Five Years, FY 1991-1995*, pp. 65-73.

15. Ibid., pp. 67-69.

16. The ELSI acronym is sometimes applied to the programs at both the NIH and the DOE, though in most cases it refers to the Ethical, Legal, and Social Implications Program at the NIH, under the NCHGR.

17. Various schemes have been proposed in addition to those reviewed here. See, for example, Annas and Elias, "The Major Social Policy Issues Raised by the Human Genome Project," in Gene Mapping: Using Law and Ethics as Guides, pp. 3-17.

18. National Institutes of Health and US Department of Energy, *Understanding Our Genetic Inheritance: The U.S. Human Genome Project: The First Five Years, FY 1991-1995*, p. 20.

19. Ibid.

20. The distinction between sources of information is important to note. ELSI's work is focused primarily on the implications expected of HGP's applications, and the project as such is not seen as particularly problematic from an ethical, legal, and social perspective. This is not to argue that the basic research information is viewed in value-neutral terms, but it is to suggest that ELSI's mission has been operationalized in such a way as to suggest this. Arthur Caplan notes the importance of this distinction between sources as well, though it is not clear whether he is critiquing it or merely describing it. See Arthur Caplan, "Mapping Morality: Ethics and the Human Genome Project," in *If I*

Were a Rich Man Could I Buy a Pancreas?, Arthur Caplan, (Bloomington: Indiana University Press, 1992), pp. 118-142.

21. See Mark L. Pearson and Dieter Soll, "The Human Genome Project: A Paradigm for Information Management," *The FASEB Journal* 5, no. 1, (January 1991): 35-39, and Robert J. Robbins, "Challenges in the Human Genome Project: Progress Hinges on Resolving Database and Computational Factors," *IEEE Engineering in Medicine and Biology*, (March 1992): 25-34.

22. Cutter, et al., *Mapping and Sequencing the Human Genome: Science, Ethics, and Public Policy*, p. 12.

23. The brackets indicate where Wexler's testimony was momentarily interrupted and then reconstructed on the basis of other statements she made there and elsewhere. United States, Congress, Senate, Committee on Commerce, Science and Transportation, Subcommittee on Science, Technology and Space, *Hearing on the Human Genome Initiative and the Future of Biotechnology*, S. Hrg. 101-528, p. 97.

24. National Institutes of Health and US Department of Energy, *Understanding Our Genetic Inheritance: The U.S. Human Genome Project: The First Five Years, FY 1991-1995*, pp. 65-66.

25. By diagnose "directly," I refer to the ability of certain tests to diagnose genetic disease for individual patients at the molecular level, which will in many cases obviate the need for genetic linkage studies that require the cooperation of extended families. The ability to detect and confirm disease at the molecular level is one of the principal advances expected from HGP research, and it is this capability that will help make these tests available for routine medical use. For a discussion of this distinction and its relevance for clinical medicine, see Stylianos E. Antonarakis, "Diagnosis of Genetic Disorders at the DNA Level, *The New England Journal of Medicine* 320, no. 3 (19 January 1989): 153-163; Belinda J. F. Rossiter and C. Thomas Caskey, "Molecular Studies of Human Genetic Disease," *The FASEB Journal* 5, no. 1 (January 1991): 21-27; and, C. Thomas Caskey and Victor A. McKusick, "Medical Genetics," *JAMA* 263, no. 19 (May 16, 1990): 2654-2656.

26. Dorothy Nelkin and Laurence Tancredi, *Dangerous Diagnostics: The Social Power of Biological Information* (New York: Basic Books, 1989).

27. Watson and Juengst, "Doing Science in the Real World: The Role of Ethics, Law, and the Social Sciences in the Human Genome Project," in *Gene Mapping: Using Law and Ethics as Guides*, p. xvi. See also Juengst, "Human Genome Research and the Public Interest: Progress Notes from an American Science Policy Experiment," *American Journal of Human Genetics* 54 (1994): 121-128.

28. For an excellent analysis of the interrelatedness of these issues, especially as they are expected to impact the private insurance industry in this country, see

Moseley, Crandall, Dewar, Nye, and Ostrer, "Ethical Implications of a Complete Human Gene Map for Insurance."

29. See her statement in the epigraph of this chapter.

30. See Jan C. Heller, "The U.S. Human Genome Project: Mapping the Moral Boundaries of an Interim Ethic," unpublished paper, Emory University, 1992, p. 12.

31. In fact, I will suggest in chapters 4 and 5 that the effects of this research on future generations may require some new ethical and theological paradigms.

32. See Weatherall, *The New Genetics and Clinical Practice*, pp. 11-37.

33. This detection, however, will not alleviate certain ethical conundrums. See, for example, the moral "catch-22" described by White and Caskey in Ray White and C. Thomas Caskey, "Genetic Predisposition and the Human Genome Project: Case Illustrations of Clinical Problems," in *Gene Mapping: Using Law and Ethics as Guides*, eds. George J. Annas and Sherman Elias (New York: Oxford University Press, 1992), pp. 173-185.

34. Eric T. Juengst, "The Human Genome Project and Bioethics," *Kennedy Institute of Ethics Journal* 1, no. 1 (March 1991): 71. Emphasis added.

35. One hundred years may actually be more plausible than two hundred fifty years, on the grounds that certain disease genes will affect so very few individuals that they simply may not be explored in much depth. In truth, though, we simply cannot predict with more precision at this point.

36. Both of these applications could also be examined from the perspective of disease *prevention*, but prevention only begs the question we are trying to address here: whose good ought to be considered when trying to prevent a disease whose prevention can only be accomplished by controlling who will live and who will not live.

37. There is so voluminous a literature surrounding IVF as such that I will not address it here; rather, I will merely introduce IVF in order to illustrate some of the ways it is or may be used by clinical geneticists.

38. Karen Dawson, "Introduction," in *Embryo Experimentation*, eds. Peter Singer, Helga Kuhse, Stephen Buckle, Karen Dawson, and Pascal Kasimba (Cambridge: Cambridge University Press, 1990), pp. xiii-xv.

39. Karen Dawson, "Introduction: An Outline of Scientific Aspects of Embryo Research," in *Embryo Experimentation*, eds. Peter Singer, Helga Kuhse, Stephen Buckle, Karen Dawson, and Pascal Kasimba (Cambridge: Cambridge University Press, 1990), p. 3.

40. Dawson, "Introduction," in Embryo Experimentation, p. xv.

41. Dawson, "Introduction: An Outline of Scientific Aspects of Embryo Research," p. 4.

42. Elizabeth S. Critser, "Preimplantation Genetics: An Overview," *Archives of Pathology and Laboratory Medicine* 116 (April 1992): 383.

43. Critzer, "Preimplantation Genetics: An Overview," pp. 383-384. At least one disease, cystic fibrosis, can be detected before fertilization, however.

44. Ibid., pp. 384-386.

45. Compare, however, Andrea Bonnicksen, "Genetic Diagnosis of Human Embryos," Special Supplement, *Hastings Center Report* 22, no. 4 (July-August 1992): S5-S11. Bonnicksen suggests that the easy linkage between preimplantation diagnosis and IVF technology may mask some of the controversial features of the former.

46. Critzer, "Preimplantation Genetics: An Overview," p. 383.

47. See Critzer, "Preimplantation Genetics: An Overview," for a brief review of these advantages and limitations.

48. Center for Biologics Evaluation and Research, Food and Drug Administration, "Points to Consider in Human Somatic Cell Therapy and Gene Therapy," *Human Gene Therapy* 2 (1991): 251.

49. President's Commission for the Study of Ethical Problems in Medicine and Biomedical and Behavioral Research, *Splicing Life: A Report on the Social and Ethical Issues of Genetic Engineering with Human Beings* (Washington, DC: U.S. Government Printing Office, 1982), pp. 42-48. Anderson has also defined two other categories of genetic intervention, which he tries to place altogether outside the therapeutic purview of clinical genetics. They include what he calls "enhancement genetic engineering," which is aimed at enhancing what are broadly thought to be normal human traits, and "eugenic genetic engineering," which is aimed at modifying complex traits such as intelligence. See W. French Anderson, "Human Gene Therapy: Scientific and Ethical Considerations," *Journal of Medicine and Philosophy* 10 (1985): 275-291.

50. That is at least the intention. See, however, Marc Lappé's discussion of unintended affects on germ cells during somatic cell manipulations in Marc Lappé, "Ethical Issues in Manipulating the Human Germ Line," *The Journal of Medicine and Philosophy* 16 (1991): 621-639.

51. Somatic gene therapy would thereby affect who lives or does not live in future generations in ways similar to any non-genetic therapy. The distinguishing feature of the new germ-line genetic therapies is the degree of control it gives agents over who lives or does not live in future generations. It is this control that brings these issues into the relevant moral domain by extending human responsibility.

52. See W. French Anderson, "Reflections: Of Hope and Of Concern," *Human Gene Therapy* 2 (1991): 193-194. Two other earlier and unsuccessful attempts have been documented, one in Germany and another in the U.S. See President's Commission for the Study of Ethical Problems in Medicine and Biomedical and Behavioral Research, *Splicing Life: A Report on the Social and Ethical Issues of Genetic Engineering with Human Beings*, pp. 44-45.

53. Results reported in Sherman Elias and George J. Annas, "Somatic and Germline Gene Therapy," in *Gene Mapping: Using Law and Ethics as Guides*, eds. George J. Annas and Sherman Elias (New York: Oxford University Press, 1992), pp. 142-154.

54. United States Department of Health and Human Services, Public Health Service, National Institutes of Health, "Points to Consider in the Design and Submission of Human Somatic-cell Gene Therapy Protocols," in *Human Gene Therapy*, ed. Eve K. Nichols (Cambridge, MA: Harvard University Press, 1986), pp. 195-208.

55. On this argument, see Arthur L. Caplan, "If Gene Therapy is the Cure, What is the Disease?," in *Gene Mapping: Using Law and Ethics as Guides*, eds. George J. Annas and Sherman Elias (New York: Oxford University Press, 1992), pp. 128-141.

4

Impersonal and Person-Affecting Approaches to Value

> Just as we need thieves to catch thieves, we need impersonal principles to avoid the bad effects of impersonality.[1]
>
> Derek Parfit

> [E]xistence is not a moral predicate; to be cannot in itself be either good or bad, a subject of duty or prohibition, a right or a wrong.[2]
>
> David Heyd

In order to predict and evaluate HGP's likely effects on future generations, I suggested in the introductory chapter that it would be necessary to broaden the relevant moral domain under consideration. This effort has involved a consideration in chapter 2 of HGP as a biomedical research policy, and an exploration in chapter 3 of a likely three-phase future allocation pattern resulting from HGP's research and development efforts. In the process, we uncovered two basic ways that future generations are most likely to be affected by HGP: first, by an unequal allocation among the three phases of its projected future of harms and benefits; and second, by its involvement through at least two clinical applications in what I call the problem of contingent future persons. This latter problem arises with the novel choices HGP's applications makes possible. These choices will determine to an extent never before possible which future persons will live and what quality of life they will enjoy or bear, but it is not clear on what moral basis these choices should be made.

I now want to begin our exploration of two basic ways to address the problem of contingent future persons. For reasons that will become clear as we progress, we must treat this problem before we can address the allocation pattern. The problem of contingent future persons requires us to broaden the moral domain in the third way I mentioned in chapter 1. We must consider and choose between two fundamental and competing conceptions of value. The analysis of HGP's overall justification in chapters 2 and 3 set the stage for this consideration and choice. HGP's justification is basically consequentialist in structure, and thus any evaluation of the project and its effects will require some acquaintance with consequentialism.[3] Moreover, we will see that the problem of contingent future persons is posed largely as a problem of consequentialist moral theory, and the ethical qualifications we develop below to address the problem will be qualifications common to many contemporary discussions of consequentialism.[4]

A CONSEQUENTIALIST CONTEXT

Recall that in deciding whether to authorize and fund HGP, Congress adopted an approach that permitted human genome research to go forward in a "dedicated" or "big science" form. With this approach, Congress was concerned to create or to foster the *conditions* for a future state of affairs which are described primarily in terms of the new clinical applications expected from HGP and their concomitant economic benefits. We also saw that Congress is willing to incur some potentially significant costs to our society and to risk harming at least some future individuals or groups of individuals in pursuit of this state of affairs. This congressional justification takes a rather straight-forward consequentialist form.

The fundamental aim of any form of consequentialism is that outcomes be as good as possible.[5] These outcomes can result from individual actions or the collective policies that authorize individual actions.[6] As a moral theory, consequentialism assesses the rightness of particular actions or policies on the basis of some view of the good or value that the actions or policies will actually, will probably, or are intended to bring about in the world.[7] It is thus dependent on empirical judgments about *which* actions or policies best bring about, or best maximize the probability of bringing about, the good or value in question.[8] I believe this dependency can be translated into at least two formal criteria for evaluating particular actions or the policies that authorize particular actions. We must know, first, what is considered good or valuable in order to know how to evaluate the action or policy in question and, second, we must know that the action or policy in question will actually or will probably promote this good or value.[9] (I ask the question, good for whom, below.) The review of HGP's justification in chapter 2 and its projected three-phase future in chapter 3 suggests that Congress devoted most of its effort to the second, or what I call the "strategic," criterion.

In concrete terms, the advocates behind HGP did not claim (simply) that the project would lead to beneficial new knowledge and commercially useful applications; this new knowledge and these applications were *already* being developed through traditional research mechanisms. Rather, they claimed that HGP's "dedicated" shape would in fact permit researchers to discover this new knowledge and to develop these applications more *efficiently* than those funding mechanisms traditional to biological and biomedical research. Thus, the debate behind HGP, particularly in its early years, was primarily a debate about strategy or means, and it was backed (perhaps poorly) by a rather short-range cost-benefit analysis that compared only direct opportunity costs. It was later in the debate, after the basic shape of the project was set, that questions were raised about the social costs of this project, many of which will result from the maximization of efficiency.

The concerns associated with these long-term costs and risks raise a question I now wish to explore in more depth: how can *constraints* be justified within HGP's overall consequentialist justification? In concrete terms, this question has been delegated largely to the ELSI Program. In abstract terms, the question of constraints moves us back to the first formal criterion of consequentialist justification: the view of good or value which undergirds it. It is thus a concern that logically cannot be addressed in consequentialist terms alone; other criteria must be brought to bear, for we are not trying to decide a right course of action but a theory of good or value that tells us how to determine a right course of action.[10]

Justifying Constraints

The moral leverage for any constraint in a consequentialist approach must come from the first and logically prior criterion, its theory of good or value. Thus, if HGP's advocates are to justify ELSI's constraints, they must do more than simply consider the most *efficient means* to advance this project. I would judge that ELSI's researchers are fully aware of this. However, it has long been observed that those consequentialist justifications devoted to maximizing whatever is considered of value are difficult and, in some cases, impossible to constrain. Thus, we must qualify our consequentialism in certain ways, and these qualifications will have implications below for our treatment of the problem of contingent future persons.

The first qualification we must make is between monistic and pluralistic theories of value. Simply said, monistic theories are not subject to internal constraint. This was John Rawls' principal objection to classical utilitarianism, the dominant form of consequentialism in our culture, and I believe it is valid.[11] But consequentialism may incorporate a pluralistic theory of value, and thereby provide internal constraints on its maximizing tendencies. Thus, multiple goods or values may be ranked in some order or weighed against each other on the basis of other substantive criteria. However, pluralistic theories of value encounter problems that monistic theories do not. When we compare conflicting or incommensurate goods or values, questions arise concerning the *relative* ranking or weighing of goods that may be difficult to resolve, especially in the pluralistic context of contemporary Western societies. Judgments must often be made in these situations of conflict in any case, but the concomitant ethical and empirical uncertainties increase in the process. Moreover, when the conflicted goods or values at stake are politically important, decisions are often resolved in political terms alone, as is illustrated by the NIH-DOE joint agreement. In any case, I believe we can infer from the review in chapter 2 of HGP's congressional authorization and funding that its justification in fact assumes a pluralistic theory of value, complete with a rough ranking of sorts.

The primary value at stake with HGP is ultimately—and obviously, I believe—human health. The new knowledge expected of HGP, and the clinical applications it is expected to yield, are themselves valued instrumentally for how they are thought to promote human health. Of course, whether genetic research is the best means to advance human health is another question altogether, one that Congress did not ask. One commentator observes:

> At the risk of dashing fanciful hopes, it is important to keep in mind that...[c]ystic fibrosis kills on the order of 500 Americans per year; death rates from Huntington disease are roughly comparable. These are not insignificant numbers, but they must be put into perspective. Cigarette smoking alone, for example, is estimated to cause the deaths of some 400,000 Americans per year—more, in other words, than all known genetic diseases combined. Heart disease takes an even higher toll...If improved health is our goal, then surely there is something wrong with the priorities of medical funding.[12]

Arguments can be offered to counter such claims. For example, the claim that HGP will lead to an entirely new paradigm for treating disease generally implies that this research will have applications well beyond clinical genetics. Moreover, recall that in terms of sheer cost, genetic disease alone is projected soon to consume entire health care budgets in many Western countries,[13] and unless we are simply going to ignore the increasing number of patients who suffer from genetic diseases, these costs (to say nothing of compassion) suggest that we develop and use the means available to help these patients. Also, the largely non-voluntary nature of genetic disease must be compared to the largely voluntary nature of diseases that result from smoking. People choose to smoke, at least initially; they do not choose to inherit a genetic disease.

In any case, we observed that Congress did not in fact argue that HGP is the most efficient means to promote human health generally, but merely that it is most efficient way to promote human health *through genetic means*. Moreover, it was acknowledged that the other expected benefits of this new knowledge, for example, those that advance our understanding of human evolution or help the DOE track human mutations, would not be sufficient in themselves to justify this massive undertaking. This said, even the promotion of human health was not thought to be a sufficient justification for HGP; this good was qualified by economic and political considerations.

Thus, we saw that an important secondary good undergirding HGP is its contribution to overall US economic competitiveness. As a constraint, this implies that the promotion of human health through genetic means will not be pursued at any cost. It will be constrained by allocative judgments about the most efficient means to advance this research. Additionally, we can speculate that certain possible applications will not be developed due to economic constraints; there may, for instance, be too few persons affected by some genetic disorders to justify the costs of developing applications to treat them. Lastly, these concerns were also constrained by what was deemed politically possible at the end of the Cold War, in an increasingly tight research funding environment.

There are yet other goods or values which are at stake in the decision to authorize and fund this project. These are represented by the social costs that ELSI has identified, the indirect institutional costs and the increased risk of harms to individuals or groups of individuals stemming from HGP's applications. These represent other types of goods for which policy makers must account, though they also serve to illustrate the problem with pluralistic theories of value. The sheer scale of this undertaking, in order to maximize efficiency, is a principal source of the long-term indirect costs associated with the massive amounts of information produced by HGP's diagnostic applications (in phase two of its likely future). But protecting individuals from the adverse consequences of the predicted harms stemming from the diagnostic information overload problem will raise HGP's indirect economic costs and, as a result, lower its overall efficiency. Said differently, these goods conflict. Just how deeply these indirect costs will cut into the overall efficiency of the project is an interesting technical question, but I leave it to medical economists.

I now want to turn to the question of whose good this project is ultimately intended to serve. This question must also be addressed through a theory of value; however, since it is complicated by the problem of contingent future persons, we must move to a more basic level of consideration. Here we must choose between two ways of conceiving how value is fundamentally connected to the world. To understand why this is the case, we must take an extended and rather complex tour of the work of two moral philosophers, Derek Parfit and David Heyd.

THE NATURE OF VALUE

David Heyd claims that the moral challenge of contingent future persons forces us to choose between two fundamental approaches to value, both of which are "mutually exclusive and exhaustive alternatives."[14] He puts this choice to us in the following way. We can *include* contingent future persons in the moral domain, but only by appealing to a rather counter-intuitive *impersonal* approach to value; or, we can stay with a common-sense *person-affecting* or *person-oriented* approach to value, but then we must exclude contingent future persons from the moral domain and consider them only *indirectly*, by which he means we can consider only the actual or probable effects their existence or non-existence will have on other persons. The first option is developed by Parfit and the second by Heyd himself.

At stake in this choice between two approaches to value is another choice between two ways of conceiving how value, and thus ethics, are fundamentally "connected" or related to the world.

> [The crucial difference between the two approaches] consists in the *sort of conditions that must be satisfied for a value judgment to be applicable*: certain absolute characteristics of the world as against certain person-relative states of the world.[15]

An impersonal approach views value as attached to the world or as a property of the world. It thus "characterizes states of affairs in the widest global meaning of the term."[16] This is not to imply that an impersonal approach to value will treat persons indifferently, though it does significantly qualify the moral status of persons relative to the larger, non-personal universe. Particular persons who might be affected by our actions and policies are not accorded the sovereign status they often enjoy in person-affecting approaches of value, and thus they are not available to "trump" those choices that involve contingent future persons. Rather, choices are justified ("impersonally") by reference to the overall quality and quantity of value which contingent future persons can be expected to contribute to the world when and if they come into existence in certain numbers and with certain identities.

In contrast, a person-affecting approach to value views value as something that is attached only to persons, "through the way it *affects*...subjects."[17] In this view, the existence of a valuer is a necessary condition for the existence of value.[18] Moreover, Heyd does not hesitate to carry this distinction to its logical end. In affirming a person-affecting approach to value, Heyd argues that the world in itself has absolutely no value; rather, "human beings *invest* a valueless world with value."[19] Heyd's approach to value is thus not only person-affecting in a formal sense, it is utterly anthropocentric.

Heyd uses the term "anthropocentric" to imply that humans are the only persons that ought to count in his person-affecting approach of value. We will see, however, that this approach can logically refer to God as well (insofar as God can be construed as a "person"). I pursue this possibility below, suggesting that a "theocentric" approach to value may open avenues for addressing the problem of contingent future persons that Heyd does not develop. But for these avenues to be realized, I will need to develop a critique of anthropocentrism. For, with Parfit, though for different reasons, I believe anthropocentrism is something that must be overcome if we are to accord both contingent and non-contingent future generations their due. Now, however, I want to explore in some depth the implications of these two approaches to value for addressing the problem of contingent future persons. Parfit's and Heyd's positions are representative of the two fundamental options that are available to us, and these options provide the key categories for a theological consideration of the problem in the next chapter.

An Impersonal Approach to Value

Parfit develops his argument for an impersonal approach to value, which he calls a search for "Theory X,"[20] at the end of a long and complexly qualified defense of consequentialism and of certain Buddhist-like views of persons. I will not review Parfit's views of persons here. They lend some credence to an impersonal approach to value, but by his own admission they are not decisive. One could rationally adopt an impersonal approach to value without believing that persons ought to be construed in Buddhist-like ways, and the converse is true as well. I will, however, review key parts of his defense of consequentialism, for a number of his qualifications are important below for understanding how he evaluates the effects of far-reaching policies and, in particular, the problem of contingent future persons. In essence, we must understand how a consequentialist can defend him- or herself against charges of being overly individualistic, and we must clarify the different ways the term "impersonal" can be used to qualify an approach to value.[21] In the process, we will also learn how one moral philosopher develops a non-anthropocentric approach to value.

ADDITIONAL QUALIFICATIONS

Critics often charge that consequentialism is too individualistic and overly concerned with the actual effects of individual actions. Indeed, some consequentialists conceive of their moral theory in this way.[22] But Parfit believes this understanding of consequentialism is both unpersuasive and unnecessarily demanding, especially when it is applied to what are typically called "coordination problems." These problems arise for agents whose actions either affect or are affected by other agents, especially on the level of public policy formation. With coordination problems, the individual agent must be concerned with what makes outcomes best, given what others will actually do. Parfit believes this places an almost impossible empirical burden on the prediction and control of outcomes. For instance, if this individualistic form of consequentialism were the only one available to us, we might in principle be unable to evaluate the effects of HGP on future generations. There are simply too many future intervening agents with which to contend. To counter such problems, Parfit proposes that consequentialism be qualified in a number of ways.

We may, for instance, distinguish between objective and subjective right and wrong. An objectively right act is the one agents have the most reason to do (which they might know if they had god-like foresight or perfect hindsight). A subjectively right act is the one agents have most reason to do given what they believe, or ought to believe, *at the time a choice is made.* This distinction is particularly important when making prospective choices that are correlated to an array of probable long-range outcomes. Thus, when we are evaluating a "risky" course of action (one that has outcomes about which we are empirically uncertain), we ought choose that course of action that has the greatest *expected* goodness (that is, greatest expected goodness as understood subjectively).[23] This qualification does not, of course, imply that the choice could not be altered at a later time when new information comes to light (though, such reviews increase information costs).

Parfit further qualifies his consequentialism by adopting what he calls a "collective" form of it, one that is primarily concerned with ideal outcomes. This form of consequentialism claims that each of us ought to do that which, if *everyone* did this, the outcome would be best.[24] Such a position permits *non-optimal* outcomes for individual actions if the relevant conditions are satisfied. For example, a form of individual consequentialism might obligate us to give our incomes to feed the poor, simply on the grounds that this money would actually maximize good outcomes for those poor people. However, following a collective form of consequentialism might obligate us only to give some portion of our incomes (perhaps in taxes), along with all other citizens, to help these same poor people. Thus, collective consequentialism helps lessen the empirical burdens on agents and bring consequentialism more closely in line with our long-held moral beliefs. For Parfit, however, this qualification also provides a crucial defense against certain types of theoretical failure that worry him.

Parfit believes that moral theories can "fail" in a number of ways, and he believes that some forms of failure are more serious than others. The problem of contingent future persons is at least a *prima facie* case of one type of theoretical failure. The fact that common-sense notions of harm and benefit do not seem to apply to those actions on which the existence of future persons is perplexing.[25] Parfit's defense of consequentialism represents ways that it can be qualified to address the challenges of such problems while, at the same time, it does so in such a way as to imply many of our long-held common-sense moral beliefs. But Parfit is also concerned that moral theories can fail in different way, that is, by being *self-defeating*. This kind of failure requires no competing assumptions from other conceptions of morality, and thus it is all the more serious a failure. Under this type of theoretical failure, moral theories give aims to agents (individual and collective) such that, if the agents actually achieved them, their success would work to undermine these aims. An example of this type of failure can be developed from the case in the previous paragraph.

Again, following the guidance of individual consequentialism we might actually reduce the suffering of poor people in the world, at least for a short time, by giving them much of our income; however, in doing so we might inadvertently increase the overall number of poor in the world, thereby undermining the aim of the theory to reduce the suffering of the poor. For instance, it could be the case that giving our incomes to the poor might leave our own families destitute in the process, and leave us unable to feed them. Moreover, if *everybody* gave their incomes to help the poor, the poor might eat for a few days but the world might soon be a *much* worse place for many more people. However, if we follow a collective form of consequentialism, we might instead give a small portion of our incomes to alleviate the suffering of the poor, and keep the major portion to feed our families and to invest, thereby helping to create additional wealth that could help many more poor people overall.[26]

In fact, Parfit believes that moral theories can be self-defeating in four (4) ways. We need not review these here, except to note that he believes it is much worse for a moral theory to fail collectively than individually. If I follow the guidance of an individual form of consequentialism and give my income to the poor, the bad effects of this act are very limited. However, if all citizens follow the guidance of a collective form of consequentialism, and *it* fails, the results could be catastrophic literally for millions of people. Below, as we review the problems associated with addressing the problem of contingent future people, we will continually move back and forth between the individual and collective levels. A failure of a moral argument on the individual level, relative to the problem, will be a matter for concern, but a failure on the collective level may be grounds for rejecting the argument altogether.

Next, and perhaps most importantly for our discussion below, Parfit claims that consequentialism ought to be agent-neutral.[27] By agent-neutral, Parfit means that all agents are given common aims. This position is contrasted with agent-relative moralities, which permit agents to adopt different aims. For example, with an agent-relative morality, an agent is not permitted to torture an innocent person even if by doing so he or she would surely save many other innocent people from being tortured. Parfit claims that the aim of agent-relative moralities in such cases is to keep the *agent* from committing a moral wrong, regardless of the regrettable outcomes for other innocent persons. On the other hand, the aim of an agent-neutral morality in such cases is simply that there be less torture or suffering, and this aim is given impartially to all agents. Thus, for the agent, the question of whether to participate in the torture becomes a question of strategy or means: is it reasonable to believe that the agent's participation in the torture of innocents will save other innocent people from being tortured?

This distinction is important to the discussion that follows because it can be a source of confusion for understanding what constitutes an impersonal approach to value. An agent-neutral morality represents one way an ethic can be "impersonal."[28] Thus, using an agent-neutral morality in the case of innocents being tortured, the agent would be morally required to do the same act regardless of his or her personal relationship to the innocents. In this sense, then, the term refers to the *perspective* one ought to adopt when justifying particular actions or policies and *not* to the impersonal approach to value as such. An impersonal approach to value refers to the *nature* of value, in the sense that value is attached to the world as a state of affairs or that value is attached only to persons (our only two options). Below I will use this distinction between perspective and nature to argue that it is possible to develop a theocentric ethic that is agent-neutral with respect to the perspective an agent ought to adopt when making choices (and thus "impersonal" in the sense of impartial), and person-affecting with respect to its approach to the nature of value. (Parfit, of course, would not be persuaded to follow such an ethic, but I believe he would agree that it is analytically possible.)

Parfit makes additional qualifications to his understanding of consequentialism as he addresses the problem of contingent future persons. Primarily, he develops an impersonal approach to value. However, by arguing for this approach, Parfit is not attempting to mount an argument for a substantive approach; rather, he is trying to qualify any approach that might be chosen.[29] Any theory of value that we use, he claims, ought to be impersonal.

NON-IDENTITY PROBLEM

The case that leads Parfit to conclude we need an impersonal approach to value concerns what he calls the "problem of non-identity." This is Parfit's perplexing designation for the problem of contingent future persons (others have renamed it the "Parfit problem" or the "identity problem").[30] If we care about the outcomes of our actions or policies for future generations, Parfit argues that there are at least two *types* of outcomes that ought to concern us. We can affect the *identities* of individuals who will live in the future and we can affect their *numbers* (Heyd adds a third category, existence, which Parfit subsumes in identity). These two types of outcomes give rise to an array of three types of choices relative to these kinds of far-reaching actions or policies:

Same people choices, in which neither the identities nor the number of future people is affected by the choice;

Different people, same number choices, in which the choice will lead to the same overall number of future people, but to people with identities different from those who would have existed without the choice or with another choice; and,

Different people, different number choices, in which the choice leads both to different numbers of people and to people with different identities.[31]

Same People Choices

"Same people choices" is Parfit's awkward designation for categorizing actions or policies that can meaningfully be said to harm or benefit future people. By "same people" Parfit means that the same future people will live regardless of the choice, action, or policy in question. These are people whom I call "non-contingent future persons" and whom Heyd groups together with currently living people and calls "actual" people. Parfit does not believe we ought to use a person-affecting approach to value to evaluate such choices, but he acknowledges that non-contingent future persons can be accounted for in person-affecting terms. For example, the future generations affected by HGP's projected allocative pattern can be treated under this type of choice, as the identities and numbers of the future persons affected by the allocation are not contingent on HGP in the relevant sense. Thus, if we could be sure that HGP would not be involved with the problem of contingent future persons, we could treat allocation pattern in terms common to biomedical ethics generally. That is, it would be meaningful to speak of these future people as being harmed or benefited by HGP's applications.

With respect to these "same people choices," Parfit believes that their distance from us in time is not morally relevant. To make this point, Parfit uses an analogy between those persons who are distant from us in space and those who are distant from us in time. Parfit believes that the relative distance or closeness to us of the persons affected by our actions is morally irrelevant. Analogously, he argues, the distance in time from us of those future persons affected by our actions is no more relevant than distance in space is for actions that affect persons contemporary to us. These considerations lead Parfit to yet another qualification of his consequentialism. He illustrates the point by appealing to some simple casuistry. Suppose, he says, some individual drops a piece of broken glass in the woods and one hundred years later a child steps on it and injures her foot. Though we cannot accurately predict such effects with the empirical tools at our disposal, this does not imply (for Parfit) that they should be morally ignored. The child will be a person who could be harmed by this act, even though she does not exist now. The glass should have been safely buried as a precaution.[32]

Similarly, on the collective level we should take account of possible harms to future people. The fact that some of these harms may occur thousands of years from now is, again, morally irrelevant to Parfit. He thus cannot justify the use of a social discount rate when making policy choices in cost-benefit terms. To illustrate this point, Parfit appeals to another case, the disposal of nuclear waste. At a discount rate of only five percent per year, Parfit claims that one death next year due to nuclear waste contamination would count for more than a billion deaths in five hundred years. We can be relatively confident that future people will exist in five hundred years and we can be absolutely certain that the nuclear waste material will be highly dangerous to them. According to Parfit, policies that neglect such long-range consequences on the assumption of a social discount rate are morally "indefensible." He allows that discounting may be justified in economic terms—remoteness in time, Parfit says, may indeed be related to such economically important facts as predictability[33]—but this provides no moral justification for disregarding the effects of our policies on future generations. Again, remoteness in time is as morally irrelevant as remoteness in space.

The conclusion of this discussion for Parfit is that any approach of value, impersonal or person-affecting, used to account for our actions relative to future generations must be *time-neutral*. This means that the interests, rights, or welfare of future generations are not discounted simply because they live in the future (nor, however, are they given any more weight than currently living generations). Again, time may legitimately count as an empirical variable in predicting and controlling outcomes, but it should not count morally in evaluating them.[34] By not counting morally, I believe Parfit is arguing that even though we cannot predict that particular persons will be harmed by our actions, the fact that some could be harmed is enough to justify precautions relative to their interests or welfare. In this respect, then, he is "risk-averse" with respect to future generations.

Time neutrality thus has implications for addressing HGP's effects on non-contingent future generations as well, especially when it is coupled with those qualifications outlined above. The primary problem we face in this case is empirical, that is, it is one of predicting HGP's effects of future generations accurately enough to base our evaluations on them. Parfit's qualifications, however, limit our need to predict these effects with absolute accuracy. Rather, taken together, they suggest that what needed is a reasonable judgment concerning probable benefits, costs and harms at the time the decision is made, subject to periodic review as the research progresses (which HGP is). Further, the time-neutral qualification implies both that it is morally irrelevant that HGP's phase two costs and harms will be born by persons who do not now exist, and that these costs and harms should not be discounted simply on the assumption that researchers will be successful in developing future therapies in phase three. But before we pursue this further, we need to see how Parfit evaluates those more problematic actions on which the identities and numbers of future persons are contingent.

Different People, Same Number Choices

Having made an argument that future persons cannot morally be ignored simply because they will exist in the future, Parfit observes that people who are remote from us in time are in fact different from those who are remote from us in space in one morally relevant respect. The identities and numbers of those who are remote in time can be affected by our choices. This observation permits Parfit to introduce what he calls the non-identity problem, which he does in emotionally powerful terms. He observes that:

> Each of us might never have existed. What would have made this true? The answer produces a problem that most of us overlook.[35]

The problem most of us overlook is the contingency of our own existence and the ethical implications this has for evaluating actions and policies. These implications were always present for us in one sense, but they are rendered morally relevant only recently because of our increasing knowledge of human reproduction. Parfit's discussion also illustrates how the problem of contingent future persons can arise in ways that have nothing to do with the elaborate technologies being developed through HGP.

On the basis of our understanding of reproduction, Parfit argues that our identities are time-dependent, and by this he means that our identities are dependent on the *time of our conception*.[36] If a different sperm cell had joined with either the same or a different egg, we simply would not have existed. Thus, at any given period in our parents' lives, the chance combining of different sperm and egg cells could have resulted in a *different* person being born in our stead; hence, a "different people" choice. The same outcome could have occurred if our parents decided to postpone conception for a year or even a month.[37]

These biological facts are of interest to Parfit on both the individual and the collective levels. They give rise to the problem of contingent future persons on the individual level when Parfit is trying to decide on what grounds a teenage girl should postpone a possible pregnancy. Intuitively, we might be tempted to ground our reasons in the harm her child would experience if she gets pregnant and gives birth to it at such a young age—Parfit refers to this harm as a "bad start" in life. Parfit observes, however, that we cannot base our arguments on the harm her future child might experience because, if she delays her pregnancy, a *different* child will in fact be born.

> What [then] is the objection to her decision [to conceive a child that will have a bad start in life]? The question arises because in different outcomes, different people would be born.[38]

What we need in cases like this, says Parfit, is an impersonal approach to value, one that can meaningfully address the question of the quality of life of the future child, *regardless of who is born*. Said differently, the identity of the particular future child that is born as a result of the mother's choice is morally irrelevant to this conception of value; what matters morally is the quality of life the future child will enjoy. On this basis, then, we could argue that the teenage mother ought to delay her pregnancy because her child would then have a higher probability for a better quality of life. With an impersonal approach to value we are trying to maximize the quality of life in the world generally rather than the quality of a particular person's life, on the assumption that this would make the world a better place overall. Interesting, then, this approach to value renders both of the contingent future children this teenage mother could have—both the one that would be born if she does not delay her pregnancy and the one that would be born if she does delay it—morally relevant to her decision. They are both, thereby, included in the relevant moral domain, even though neither exists yet and at least one will never exist. This approach to value thus solves the problem of contingent future persons on the individual level.

This said, Parfit is even more interested in the collective implications of the non-identity problem. He also observes that we might not have existed if our parents were affected by the choice of a particular social policy that led them to delay conception, to abort a conceived fetus, or to avoid conception altogether. These choices likewise affect who will live and who does not live, and justifying such choices again presents special ethical challenges.

For example, Parfit says, suppose we are trying to choose between one policy that will conserve natural resources and another policy that will radically deplete them. Is there any *moral* reason we should choose the conservation policy over the depletion policy?[39] Again, in response to this question, one might intuitively suggest that the depletion policy will obviously harm future persons in some way and that, other things being equal, this is a sufficient reason to choose the conservation policy. Parfit makes this argument himself above when discussing the case of the future child injured by a piece of glass left in the woods. But, in that case, the child's existence was not contingent on the choice in the relevant sense. In this case, however, the alternative policies under consideration lead to *different* populations of people. If the conservation policy is chosen, a chain of events can reasonably be expected to unfold such that the future generations who are born under such a policy will have different identities than the population of those born under the depletion policy.[40] How should we evaluate such an outcome?

Parfit believes that if we were to ask the future people who might exist under either choice whether their lives are worth living,[41] they would almost surely answer affirmatively. That is, even the future generations who would live under conditions of radically depleted resources are likely to find their lives worth living, and thus they are not likely to wish they had never been born or to commit suicide. Moreover, Parfit believes that once we explain to them that they would not be alive if the depletion policy had not been chosen, it is unlikely they would object to its choice, however much they might regret their circumstances.

The problem of contingent future persons—the so-called non-identity problem—arises on a collective level with this kind of case. It says that if we are concerned with the effects a policy choice will have on *particular* persons, and if our policy is one that affects the timing of conception for these future people, we will have no *moral* reason not to choose the depletion policy over the conservation policy. The depletion policy will *not be worse* for those who will live in future generations; that is, we cannot meaningfully claim that these particular persons have been harmed by the choice of the depletion policy. If the depletion policy is chosen, a different set of people will live and, if their lives are even marginally worth living, it is hard to imagine they would wish themselves not to have existed.

This said, Parfit believes that the deliberate choice of the depletion policy is morally wrong. However, in this case, his intuition of moral wrong logically cannot be justified by reference to the future persons who are in fact born as an indirect result of the policy's choice; rather, he must employ an approach to value that is *impersonal*. An impersonal approach to value, Parfit claims, "is not about what would be good or bad for those people whom our acts affect."[42] Rather, it is concerned with states of affairs in the world. On this basis it is possible to argue that the choice of the depletion policy is wrong because it leads to worse states of affair in the world, again, *regardless of who is born as a result of the choice*. Particular persons are not treated indifferently in this approach, which is to say that persons enjoying a full life without the fears of a radically depleted environment represent a better state of affairs than persons living marginal lives in a radically depleted world. However, these persons are not available as a moral "trump" when justifying actions or policy that affect future generations in the relevant senses.

Now if Parfit had stopped here, our choice between theories of value might be easier (and this chapter would be shorter). However, adopting an impersonal approach to value raises other problems for Parfit as he turns from "different people, same number" choices to "different people, different number" choices. In short, Parfit cannot find ways theoretically to constrain an impersonal approach to value, and it leads him to what he calls the "repugnant conclusion," the "absurd conclusion," and to various forms of what he calls the "mere addition paradox." These colorful terms should be clearer to the reader momentarily.

Different People, Different Number Choices

Some actions or policies will not only affect who is born, they will also affect how many people are born. If our choices affect the number of people in the world, then Parfit believes we must eventually confront the question of how many future people there should be, that is, how many there should be ever. Such hypothetical questions are intended to test the rational and moral limits of Parfit's impersonal approach to value. In this case, we are brought back to the problem of scarcity, mentioned above when discussing cost-benefit analysis. We can predict that a threshold exists beyond which any growth in population will begin to lower the quality of life for given populations. The question Parfit asks here is whether a loss in the quality of life by increasing the number of people in the world is morally outweighed by a sufficient gain in the quantity of people in the world. He illustrates this problem with examples from the utilitarian tradition; his own consequentialism, which is not strictly utilitarian, is subject to similar concerns.

Classical or total utilitarians argue that considerations of quantity ought to constrain quality. Conversely, ideal and average utilitarians argue that quality ought to constrain quantity. Consider first the problem of trying to constrain quantity with some conception of quality. Imagine, suggests Parfit, a world "A" in which a given population enjoys a relatively high quality of life. Now imagine a different world "B" in which the population has doubled and the quality of life has been slightly lowered for all people. More people exist in B, and all of them find their lives worth living. Many of us would find the B world morally acceptable, or perhaps even preferable to A. But Parfit observes that total utilitarians are *obligated* to pursue or to promote the B world.[43] They ought to maximize whatever is of value, and the principle of value is understood in quantitative terms. Parfit's so-called "repugnant conclusion" results from pursuing this kind of case to its logical end.

At the limit, we can project a "Z" world in which the quantity of people has grown to such an extent and the quality of their lives has been reduced to such an extent that people's lives are barely worth living. The thought that we are morally obligated to produce this kind of Z world is intrinsically repugnant to Parfit.[44] Yet it is implied by both the person-affecting and impersonal total versions of utilitarianism. Thus, Parfit argues, "If we are convinced that Z is worse than A, we have strong grounds for resisting principles which imply that Z is better."[45]

Parfit then asks whether an adequate constraint could be devised that would limit this maximizing process. He believes that it might be plausible to place an upper limit on the *positive* value of quantity beyond which we are not obligated to increase the numbers of people who live, but that it is not plausible to place such a limit on the *negative* value of suffering. Here Parfit is in agreement with the critics of classical utilitarianism. He does not believe the suffering of a few individuals could ever be outweighed by the happiness of the many. But the logic of this constraint leads him to the "absurd conclusion" when the population grows beyond a certain threshold.

> Badness that is unlimited [finally] must be able to outweigh limited goodness. If [a] population was sufficiently large, its existence would be intrinsically bad. It would be better if, during this period, no one existed.[46]

Parfit, of course, believes the existence of the human race should be valued.[47] So, to avoid the absurd conclusion, he must reject the constraint that follows from the asymmetry of the argument concerning the upper limit of quantity.[48]

Parfit then investigates an alternative way to constrain quantity by permitting the quality of life to override quantity. In this case, the average form of utilitarianism is tested and yet another set of problems arises, which Parfit groups under the "mere addition paradox." These problems are somewhat more complicated to explain. Their designation is meant to imply that merely adding people (to maximize value) results in a paradox when trying to constrain such actions with some conception of the quality of life. On the average principle, it is worse if there is a lower quality of life, per life lived.[49] Thus, on this principle we are obligated to produce children only if their expected quality of life is higher than the average quality of life of existing people and, conversely, we are prohibited from bringing any child into the world if its expected quality of life will fall below the average. This principle does indeed constrain the maximization of quantity, but at considerable cost. At the limit, this position leads in the direction of what is sometimes called "negative" utilitarianism. This form of utility assigns a higher priority to preventing badness than promoting goodness. And, as life is obviously the condition for any suffering, it inevitably leads to a recommendation that the entire human race be painlessly allowed to die.

Even before the limit is reached, however, the "mere addition paradox" highlights the arbitrariness of such a standard. The paradox arises because of an indeterminacy in the "base population" that is used to determine the average quality of life. David Heyd, in criticizing Parfit, calls this the "problem of assignability,"[50] and it besets all forms of utilitarianism (and forms the basis for one of the reasons Heyd rejects an impersonal approach to value). The utilitarian cannot answer, in utilitarian terms, the question of *whose* good ought to be included when the average base-quality of life is calculated. Should we include currently living generations within a certain bounded territory or all currently living generations in the world; or, should we include future and even past generations as well? Given these possible populations, the various forms of the mere addition paradox result. They result *because we cannot evaluate outcomes consistently*. Given one base population, we evaluate an outcome favorably. Given another base population, we evaluate the same outcome unfavorably. They cannot both be true, so we are left with a paradox and no way to resolve it.

In the end, then, Parfit fails to find his "Theory X." He claims that his complexly qualified approach to value must solve the "non-identity problem" (the problem of contingent future persons), avoid both the "repugnant" and "absurd" conclusions, and solve the "mere addition paradox." Parfit's impersonal approach to value solves only the non-identity problem, though he holds out hope for a solution to the other problems which, as we saw, are primarily theoretical problems of constraint. David Heyd exploits these weaknesses against Parfit in developing a person-affecting or person-oriented approach to value to address the problem of contingent future persons. However, his position is not altogether satisfying either, though for different reasons.

A PERSON-AFFECTING APPROACH TO VALUE

David Heyd believes his person-affecting or person-oriented approach is finally a better way to conceptualize the fundamental approach to value. This approach, he says,

> holds (contrary to its impersonal rival) that value is analytically related to the needs, wants, interests and ideals of *actual* human beings and cannot be ascribed "to the world."[51]

However, in trying to justify this choice for a personaffecting approach to value, Heyd agrees with Parfit that there is no "knockdown" argument or ethical "foundation" on which to begin. Both authors simply begin with their assumptions and "cash them out" in terms of their costs and benefits relative to many long-held moral beliefs.[52] Thus, Heyd states:

> ...I do not argue with the truth or falsity of impersonalism. I only claim that we are confronted with a choice that cannot be avoided: either to adopt an impersonal view of value and pave the way for the inclusion of potential [contingent] people in the moral community, or adhere to a person-affecting view and treat all genethical choices [that is, choices on which the existence, identities, and numbers of future persons are contingent] either as a matter related only to actual [non-contingent] existing people or consider them as lying beyond the scope of ethics altogether. Only a theoretical cost-benefit analysis will settle this global dilemma...[53]

The fact that Heyd's person-affecting approach to value limits the moral domain in such a way as to exclude contingent future persons is, as I have already indicated, a matter of grave concern for me. I believe it leaves the future persons who are eventually born overly vulnerable to the interests of the agents who made the choice that determines their existence. Heyd, of course, is aware of this concern, but he believes that agents will be adequately constrained by their own interests or the interests of society generally without resort to an impersonal approach to value. Thus, while Heyd believes that contingent future persons as such lie outside the relevant moral domain, he does not ignore them altogether. For though a choice that brings contingent future persons into existence cannot meaningfully be said to harm or benefit them, such a choice could harm or benefit *other* persons who are affected by the choice. He calls these "other" persons "actual persons" and contrasts them with "potential persons," which is his term for contingent future persons. His development and use of these terms is counter-intuitive, however, and thus requires some elaboration.

Actual and Potential Persons

With the distinction between actual and potential persons, Heyd intends to convey a *moral* distinction between persons. This is in contrast to a metaphysical or a logical distinction, which is the more common way to understand these terms. By "moral," Heyd means that it answers a question that must be answered in any person-affecting or person-oriented approach to value: for whom is the good intended? For Heyd's person-affecting approach to value, the persons for whom the good is intended ought always to be—indeed, can only be—*actual* persons. Heyd defines actual persons as those persons "who do not owe their existence to a human choice."[54] Thus, for Heyd, actual people include not only all persons who presently exist, but also those who will exist in the future, *independent of our choice*. I have called these future actual persons "non-contingent persons," thereby hoping to avoid some of this confusion. This is the same category of future persons that Parfit treats under his "same people" choices. In any case, Heyd is indicating that he agrees with Parfit that we ought to include non-contingent future persons in the moral domain. The fact that these future persons do not exist yet gives us no grounds to ignore their interests, rights, or welfare when undertaking actions or policies that could affect them. In contrast to Parfit, however, Heyd believes that such persons can morally be subjected to a social discount rate.[55] That is, Heyd believes that persons in the more distant or remote future may make less moral claim on us than those in the nearer or proximate future. Nevertheless, remote or proximate, they remain subjects who are included in the moral domain and who warrant at least minimal or *prima facie* regard and consideration. Again, by definition, actual persons and only actual persons have moral standing in Heyd's person-affecting approach to value.

Conversely, potential persons are defined as those people whose existence is contingent on or dependent on human choice.[56] According to Heyd, potential persons ought to be accorded absolutely no *moral* standing. They cannot be harmed or benefited by being brought into existence, and thus they may be morally disregarded in any choice on which their existence depends. This position assumes that life itself, that coming into existence, is not a benefit or harm but merely a condition of benefit or harm.[57] Thus, when considering actions or policies that lead to the creation of such persons, we are not morally obligated to consider their alleged interests, rights, or welfare; indeed, they cannot meaningfully be said to have interests, rights, or welfare. Thus, the principal concern that emerges in addressing the problem of contingent future persons with a person-affecting approach to value is being able to determine who is the potential and who is actual; however, this proves to be very problematic as the determination is always relative to the agent or agents in question.

PURE AND IMPURE GENESIS CHOICES

Heyd correctly points out that the importance of the distinction between actual and potential persons "cannot be overstated."[58] It is the linchpin of his argument. Again, however, the fact that Heyd accords potential persons no moral standing does not imply the he ignores them altogether. Choices that involve potential or contingent future persons are called "genesis choices," which are a more generalized form of Parfit's "different people choices." Heyd claims that genesis choices involve questions of "world-creation," and may be contrasted to traditional ethics which are said to involve questions of "world-amelioration." Questions of world-creation require, Heyd believes, an entirely new field of ethics, which he calls "genethics" (meaning genesis-ethics, not gene-ethics). Genethics is defined specifically as a "field concerned with the morality of creating *people*."[59] Heyd hopes this new field will provide a

> common theoretical framework for the analysis of ethical aspects of population policies, family planning, genetic engineering, education, environmental ethics, intergenerational justice, "wrongful life" cases, and even a theology of creation.[60]

His interest in outlining a theology of creation will be pursued as a theological concern below. Here I am concerned with how Heyd uses a theological discussion to distinguish between what he calls "pure" and "impure" genesis choices.

Heyd begins with a reference to the first creation myth in Genesis.

> [I]f the world as a whole does not exist, how can its creation be considered good? Or, alternatively, if no human beings exist (in the inanimate world created in the first five days), how can their creation be considered of any value? God retrospectively judges his creation as "very good," but what kind of good is it? There are two possible responses to this question. The first is to take the *impersonal* wording of the biblical verse literally, that is to say the state of affairs after Man's creation is better than the one preceding it.... The second is to view the value of the newly created state of affairs [relative] to God, that is to say to see it as good *for him*.[61]

Heyd, of course, adopts the second approach. However, his reference to the original creation permits him to develop an ideal type of genesis choice, what Heyd calls a "pure" genesis choice. In a pure genesis choice, only one "person" exists to be affected by the choice, namely, God. His reference to the original creation allows Heyd to uncover the relevant categories of analysis for impure genesis choices: existence, numbers, and identity.

Thus, Heyd claims, when God was contemplating the creation of the first humans, God first had to decide whether to create humans at all. This decision concerns the sheer existence of humans, and the inclusion of this category permits Heyd to raise the speculative question, "Is it *good* to have human beings at all?" In a person-affecting approach to value, however, Heyd points out that there is no way to ask (impersonally) whether the world is better with or without humans. In person-affecting terms, Heyd's question can be answered relative to the only "person" that is affected by the choice, God. Pure genesis choices are, therefore, utterly agent-centric or, in this case, theocentric. Second, once God decided to create humans, the question of their numbers can be raised. This category permits Heyd to ask, "How many human beings should there be?" This is the limit question that dooms Parfit's treatment of "different people, different numbers" choices. But again, in Heyd's person-affecting approach to value, the question of numbers can only be answered relative to the interests, rights, or welfare of the one actual person for whom the choice could be good, namely, God. Finally, when creating humans God needed also to decide their identities. This category permits Heyd to ask, "What *kind* of creatures should they be?" In the original creation, it too can only be answered in terms of God's good.

Humans obviously do not face genesis choices in the same way God did, and their choices relative to potential persons are thus "impure." By impure, Heyd means that other actual persons, *in addition to the agent*, either currently exist or will exist in the future who can be affected by choices that involve potential persons. These other actual persons can make legitimate moral claims on the agents who are involved in genesis choices, and it is only as other actual persons are affected by the agent's genesis choices that the implied egoism of Heyd's genethics can be constrained. The basis for this egoism is found in the analogy he sets up between God's agency and human agency.

This analogy rests on the observation that God answered the third question, what kind of creatures should humans be, by creating humans as moral subjects. Thus, Heyd reasons, just as God is the reference point for all value in pure genesis choices, *human agents become the reference point for all value in impure genesis choices*. This explicit anthropocentrism is not a logical necessity; indeed, Heyd acknowledges that our theory of value could be theocentric. But Heyd is not developing a theological ethic, and thus with respect to his anthropocentric approach to value, he states:

> [I]f God's existence is denied, human beings are left as the only reference point or "source" of value in the world. They themselves become God, in the sense that they have full sovereignty over the existence of value. The story of creation becomes a metaphor for the human position in the [almost] pure genesis context of procreation. Human beings are forced to play

> God in a God-less world. Although they do not share the omnipotence of God, they have "dominion" over the world in the radical sense that the very continuation of the existence of any value in and of the world is dependent on their begetting valuers. In a God-less world there is no cosmic plan or transcendental design that makes humans the sovereign rulers of the earth; but without their existence as subjects for whom the earth can be of value, the earth will remain valueless.[62]

In fact, Heyd suggests that humans (figuratively) realize the "image of God" in themselves by investing the world with value, that is, by willing things and beings to have value just as God presumably willed them to have value.[63] He thus calls his anthropocentric, person-affecting approach to value a volitional theory of value or, more simply, "axiological volitionalism."[64] That which humans invest with value has value, and that which they do not invest with value has no value.

Theoretically, then, Heyd's position is relatively straight-forward. However, when treating real or hypothetical cases, his genethics can become quite complicated, and this for two reasons. First, impure genesis choices exist on a rough continuum from "almost pure" questions of procreation or existence, where the agents (usually prospective parents) are almost completely unconstrained, to "moderately impure" questions of numbers, where agents are somewhat more constrained, to "very impure" questions of identity, where agents are either unconstrained or constrained in rather complex ways that I will review momentarily.[65] Second, as already mentioned, every (impure) genesis choice depends on the relevant agent or agents being able to identify who is potential and who is actual, for it is this identification that determines who is included and who is excluded from the moral domain. However, this determination can be quite difficult on the collective level. The difficulty is partly empirical, as will see, but I believe a more problematic aspect of it arises from a qualification of genethics itself. Heyd's person-affecting approach is not only substantively anthropocentric with respect to the nature of value—or, as he variously calls it, depending on the context and the agents in question, agent-centric, evaluator-centric, or genero-centric—it is also agent-*relative* with respect to the perspective agents ought to adopt when making genesis choices. It is this latter feature of genethics that I believe finally renders it self-defeating on the collective level. These complications should be clearer as we examine Heyd's treatment of the problems encountered by Parfit.

Existence

To defend his position against Parfit's impersonal approach to value, Heyd must show both how his person-affecting approach can resolve or avoid the problems that Parfit uncovered, and that it can do so in morally and rationally defensible terms. According to Heyd, Parfit's treatment of the problem of non-identity turns on the assumption that obligations are owed only to persons whose identity is largely "fixed," by which Heyd means fixed in biological terms by the combination of a particular egg and a particular sperm cell. Allegedly, we have no duties or obligations toward future persons if certain types of actions or policies affect the moment of conception and thus the identities of the future persons actually born (these are Parfit's "different people" choices). Indeed, when Parfit's assumption is attached to far-reaching energy policies, Heyd observes that he renders nearly *all* future persons contingent or potential. This is simply not plausible to Heyd (for Parfit, of course, it is all the more reason to use an impersonal approach to value).[66] To illustrate the (alleged) implausibility of Parfit's approach, Heyd exploits his distinction between sheer existence and identity.

Suppose, Heyd suggests, that US policy makers are trying to decide on a policy that they are reasonably certain will affect Mexico in the twenty-first century. Heyd believes that the 120 million Mexicans who are projected to exist in that century must be considered as actual persons for purposes of this policy decision. It does not matter that the policy makers do not or will not know the identities of these people, and it does not matter that the identities of these people are not yet fixed. Indeed, it does not even matter that many of these future Mexicans could be considered potential persons relative to the couples who will give birth to them. Because the aggregate existence of these future Mexicans is predictable within acceptable empirical limits, *relative to these policy makers* (the agents in this case), they ought to be considered as actual persons. Thus, they can be harmed or benefited by the agent's choice, and their interests, rights, and welfare should be taken into account in any decision, choice, or action that affects them. The significant factor in this argument for Heyd is that the existence of these future Mexicans does not depend on the choice of *these* agents.[67]

Heyd then considers a case on the individual level, where a couple is deliberating about whether to conceive.[68] This is similar to the case that pushed Parfit toward an impersonal approach to value (recall his concerns with the teenage mother trying to decide whether, and on what grounds, to delay pregnancy). In Heyd's view, such a couple is concerned about a potential person. Parfit would agree; the future child is potential because its existence depends on the choice of these parents. However, in contrast to Parfit, Heyd argues that the decisions of whether and of when to conceive need not refer morally to the future child at all. Under a person-affecting approach to value, the quality of life that the potential child might or might not enjoy cannot be *morally* relevant for the couple's planning since a delay in conceiving it will result in a *different* child being born, a child who thus cannot be harmed or benefited by coming into existence. However, the couple, who are in fact actual persons, can be harmed or benefited; therefore, their interests, rights, and welfare can and ought to be taken into account morally.

This case may not seem morally problematic to some readers. But suppose that this same couple has reason to believe that a child conceived now would have a high probability of being born with a genetic or congenital defect, while one conceived a year later would have a much greater probability of being born healthy. Can Heyd provide a person-oriented reason to constrain the couple? To be consistent, Heyd must concede that the couple is not *morally* constrained to wait, even though their future child would be predictably burdened with some genetic or congenital problem throughout its entire life. In fact, he states that

> existence is not a moral predicate; to be cannot in itself be either good or bad, a subject of duty or prohibition, a right or a wrong.[69]

This said, Heyd can in fact provide person-affecting reasons for the couple to wait, and these reasons provide the only constraints on the egoism of this approach. Because the prospective parents are actual persons themselves, they have obligations both to themselves and to other actual persons as well. Thus, the potential child can be considered *indirectly* by weighing the effects its possible existence or non-existence might have on actual persons. For example, the parents may have miscalculated the costs of raising a child with an inherited defect. Also, as is often the case, these costs may be "shifted" to the wider society, and the actual persons represented by society might find these costs objectionable.[70] Since the couple and the wider society are actual persons, and since the birth of such a child could adversely affect the interests of these actual persons, these costs can be included in the moral calculus. But again, the potential child itself has no moral standing, and is left out of the decision.

In the end, Heyd believes that couples' interests in having healthy children, and that society's interests in encouraging them to have healthy children, will generally keep such couples from making choices that would in fact lower the quality of their future child's life. However, for Heyd, these are empirical not moral constraints, by which he means that it is empirically true that the vast majority of couples and societies will just choose to have healthy children if given the choice. Thus, he is forced to concede that a given couple is not *morally* constrained to make such a choice. The "natural" constraints that are in place are primarily interest-driven (except when it can be demonstrated that such choices affect other actual persons adversely), and it may of course be the case that some couples will have interests that do not conform to the majority. We will come back to this point below, when I consider once again the two cases I developed in chapter 1.

Numbers

With respect to the question of numbers of future persons, Heyd believes that Parfit is mistaken to treat nearly all future persons as potential persons. Nevertheless, Heyd concedes that many future persons must be treated as such; indeed, with the expansion of our technological powers, more future persons enter this category with each passing year. However, the thus far straight-forward reasoning of Heyd's position becomes more complex and, I believe, almost hopelessly convoluted as he moves from the only slightly impure questions of existence to the moderately impure questions of numbers and to the very impure questions of identity.[71]

We noted in our review of Parfit's position that the quantity of people who live can affect the quality of their lives. At the limits, too few or too many people will threaten the survival of a given society or perhaps the entire species, and long before this happens the quality of life for many persons will be significantly diminished. Heyd accepts this argument as an empirical observation, but argues that discussions about limits are not very useful for most practical purposes.[72] The range of possible combinations between the limits is just too great, and as long as the sheer biological survival of the species is not threatened, Heyd believes that speculations concerning an optimal population are not very useful. More to the point, however, with Heyd's person-affecting approach to value, the question of an optimal ratio of quantity and quality *simply cannot be raised* as a moral concern. There is no way to raise the abstract and impersonal question of a "best possible world" or an ideal number of people in Heyd's theory of value. Thus, on the theoretical level his approach does indeed permit him to avoid Parfit's absurd and repugnant conclusions and the various forms of the mere addition paradox (though I suspect the egoism of Heyd's genethics may be equally "repugnant" to Parfit). Practically, however, Heyd acknowledges that such questions will arise for policy makers. Policy makers may want to know the answers to some of these abstract questions about population and quality of life because they may want guidance when making allocations that will affect currently living and future *actual* persons. Unfortunately, Heyd's approach may not offer much substantive guidance to these policy makers.

For Heyd's genethics to succeed, it is absolutely imperative that the policy makers in question be able to determine who is actual and who is potential for any given choice. Heyd believes that this determination is the problem for any intergenerational policy decision using a person-affecting approach to value.[73] Heyd is certainly correct that this is an important problem; however, in trying to resolve this problem, it may be he has put the emphasis in the wrong place. He puts the emphasis on the agents' perspective, and considers situations where the agents are forced to make complex empirical determinations about who is actual and who is potential. I believe, however, that a more fundamental problem with Heyd's genethics is that the determination of who is actual and who is potential turns on the perspective of the agent or agents in question. That is, genethics is not only agent-centric or evaluator-centric with respect to the nature of value, it also seems to be agent-relative with respect to the perspective the agent ought to adopt.

Recall that agent-relativism permits agents to adopt different aims when making choices. It is not logically required of a person-affecting approach to value, but it seems to be a correlate of treating the problem of contingent future persons with a genethics that is substantively anthropocentric. Indeed, it seems to arise with Heyd's analogy between divine and human agents. Absent a deity to whom obligations are owed by human agents, an anthropocentric genethics is in fact rendered both agent-centric *and* agent-relative. This implies, then, that the agent is not only a condition for the existence of value, but the principal grounds on which a particular choice or value judgment is justified. It is this latter aspect that renders Heyd's genethics egoistic. Now, if this analysis is correct, it raises a prior question that Heyd does not address: if the determination of who is actual and who is potential turns on the perspective of the agents in question, and these agents may adopt different moral aims relative to this determination, should not Heyd's genethics be required to tell us how to determine the morally relevant agents or evaluators for a given choice? This question will not prove problematic on the individual level, where the agents are sufficiently determined, but it could be particularly problematic on the collective level.

Consider some global implications of Heyd's argument for "first world" policy makers. Since first world couples can exercise more choice in procreation (and thus more of their future children are potential), while first world poor couples and third world couples cannot or will not exercise such choice (and thus more of their future children are actual), do first world parents have an obligation to refrain from conceiving as a way of controlling population growth? Genethics would seem to suggest that we have obligations to conserve resources for the (future) actual children of poor first world and third world citizens, but no obligations to the potential children of first world citizens. But, we should not jump to this conclusion too quickly, for presumably first world policy makers also have obligations to first world couples (as actual persons) who want to have children in any case. Should the procreative liberty of first world couples be permitted to continue without constraint? Consider also the efforts by first world societies to limit population growth in undeveloped countries. Would all such efforts be unjustified on the grounds that the future children in these countries must be considered actual? Perhaps not, since these children could be potential relative to first world policy makers if their population control efforts were successful. But whose perspective should count, the poor first and third world parents or the first world policy makers?

To complicate matters even more, Heyd claims that the category of numbers cannot be isolated as neatly from questions of identity as the category of existence was isolated from both numbers and identity. We are always considering, he says, numbers of people with *particular identities* (which may be why Parfit does not distinguish existence and identity); and again, for Heyd, the relation of numbers and identity changes depending on the agent in question. For the population policy maker, Heyd claims that the identities of large numbers of people are little more than ciphers, whereas for prospective parents identity can be much more firmly fixed. Thus, again, these identities are going to be actual relative to some agents and potential relative to other agents. But my question concerns those choices that involve multiple agents, say policy makers from first world countries who are consulting with policy makers from third world countries; whose perspective is to count in such cases? That is, on what grounds can we decide who the relevant agents ought to be?

The "grounds" are finally not firm for Heyd. There is, he says, no general rule to guide deliberations by multiple agents. He thus "politicizes" the problem.[74]

> Although the generocentric [generation-centric] principle applies to these impure contexts no less than to the pure ones, the only way to protect the interests of the yet-to-be-born (actual) children and to decide who is potential and who is actual, and who is potential or actual for whom, is through a mechanism of coordinated choice....The nature of the political mechanism [controlling such a choice] is, of course, highly controversial....They [that is, these political mechanisms] belong to the difficult balancing of parental rights, the value of privacy, the extent to which a society has obligations to the consequences of procreative choices made by individuals who are protected in their choices from social interference, and so forth.[75]

It is just such coordination problems that Parfit's qualifications were intended to overcome, especially his commitment to agent-neutrality. I believe Heyd's position is finally self-defeating because he cannot tell us how to decide who is potential and who is actual in making policy decisions about numbers on the collective level. If this determination is agent-relative, there is no way to determine *morally* who ought to be the agent in controversial choices.

Identity

As a category of impure genesis choices, Heyd believes that the issues surrounding identity are among the most interesting aspects of genethics, and with this I agree. They are also among the most perplexing. Treating identity as a subject of genethics assumes of course that it is under the control of human agents, however limited that control might be. Heyd examines the genethical aspects of this control primarily in choices regarding education and genetic engineering. I am interested here in the latter.

Heyd introduces his discussion of identity by explaining how identity can be understood as a genesis problem. He explores those aspects of identity that are subject to creation, not as a once-for-all consideration as with existence, but as a gradual process of formation. He is particularly interested in this process as it is under the control of a given agent and subject to evaluation primarily *from that agent's perspective*. He must sort out the aspects of identity formation that are actual and potential, and for whom, at any given time in a person's life. As we might surmise, those aspects of identity that are subject to choice are regarded as potential, and the person who makes choices regarding his or her *own* identity is essentially unconstrained in these choices. As identity develops and certain potential aspects are actualized, the empirical and moral constraints expand, especially for "outside manipulators" of identity, such as parents, educators, or genetic researchers. But these constraints do not apply to the subject. A person-affecting approach to value views value relative to the subject *for whom* it is valuable and, on the individual level at least, there is no question concerning the identity of the primary evaluator. It is the subject him- or herself.

In terms of genetic engineering, many identity traits that are biologically or metaphysically potential must be considered actual in genethical terms. These include those traits that are not presently under human control and thus are not subject to human choice. Again, this approach leads to some rather counter-intuitive conclusions. On the individual level, for instance, it should come as no surprise that questions concerning the quality of life a given preembryo may or may not enjoy as a future person are not morally relevant to a decision to intervene genetically. Moreover, (impersonal) considerations such as the quality of the gene pool or the strengths of the natural selection process are likewise irrelevant. On the other hand, if the relevant agents value the zygote from which a harmful gene has been removed, it must be considered as at least partly actual in genethical terms and thus a moral subject in some weak sense. The key to understanding Heyd's position is, again, determining who is actual and who is potential, and for whom. For the myriad problems related to genetic engineering and therapy Heyd is the first to admit that his approach provides only the sketchiest of guidelines. He believes we simply do not have the scientific knowledge to say with much degree of certainty which genes are significant for identity formation and which are not. I do not believe this is true, however, for many of the genes that lead to known genetic diseases. Altering *these* genes will surely alter a person's future identity. In any case, as our control of inheritance increases, more traits will presumably become "potential" in the relevant sense, and thus subject to the same problems that complicate other genesis choices.

Thus, for example, if we move to the collective level, how would Heyd advise HGP researchers which genes to seek and which therapies to develop first? This choice is not without interest to affected parents or to spouses who are carriers of a known genetic disease. Suppose researchers are trying to decide between two research protocols, each of which, if successful, will permit them to treat a different genetic disease. Suppose further, however, that these researchers face funding limitations, and they can choose only one of two possible research protocols. With Parfit's impersonal approach to value, we could debate the relative value of the two protocols and their implications for the quality of life of some estimated number of potential persons. With Heyd's person-affecting approach, however, these questions are not relevant. Rather, his genethical response turns on the perspective of the agents involved in the choice, who will decide who is actual and who is potential relative to this choice. How would we make this determination? Should it be made by federal policy makers, HGP researchers, disease-oriented public interest groups, or some combination these? The most likely answer is the last one, namely, some combination of these groups. But then, how would we decide who is actual and who is potential? And who decides whether these are the groups that ought to make such choices?[76] Again, I do not believe Heyd's genethics, at least as he conceives it, can answer such questions.

SUMMARY

This chapter has reviewed, at some length, two positions that attempt to address the complex challenges arising with problem of contingent future persons. Much of the challenge and complexity is due simply to the different terminology these authors employ. Parfit treats the problem as a "problem of non-identity" under his "different people" choices; Heyd generalizes it as a "genesis choice" and treats it under an entirely new field of ethics which he calls "genethics." Parfit does not designate future persons as potential or actual, as he is indifferent to their status; what finally matters to him is knowing empirically that contingent future persons are somehow involved in a particular choice and making this choice in impersonal terms. Heyd distinguishes between "potential" persons and "actual" persons in non-standard ways, with his category of actual persons encompassing both currently living persons and future persons who will live irrespective of a given agent's choice. While I try to represent these differences in terminology faithfully, my own terminology is an attempt to clarify some of this confusion (there may be grounds on which to object to my terminology as well, but I could not think of better terms than "contingent" and "non-contingent" future persons).

These differences in terminology notwithstanding, both Parfit and Heyd share a general commitment to consequentialism, and both agree that the problem of contingent future persons (by any name) requires us to move to, and choose between, the two fundamental and competing conceptions of value. Parfit's impersonal approach to value can include contingent future persons in the moral domain and consider them directly as variables in choices where their existence and number affect the quality and quantity of life in the world. Heyd's approach, which we saw is formally person-affecting but substantively anthropocentric, excludes contingent future persons from the moral domain and thus can consider them only indirectly (by the effects of their existence or non-existence on other actual persons). Both authors also have problems with constraint, though I judge Heyd's to be more serious than Parfit's.

Now, we might find much to argue with in both of these positions, but I want to concentrate on their basic arguments in order to suggest their relevance for the next chapter's theological discussion. To do so, recall the two cases I used in chapter 1 to introduce the problem of contingent future persons.

On the individual level, both Parfit's and Heyd's arguments provide a basis for the couple who are trying to choose between preembryos, some of which are affected with a genetic disease and some of which are not. For Parfit, the choice would be made on the basis of the probable contribution *these* preembryos would make as future persons to the overall quality and quantity of life in the world. Heyd would ask the couple (and secondarily the wider society) to consider how the existence or non-existence of a potential person would affect *them*. In choices of this kind, I believe we can agree with Heyd that either approach is likely to yield similar outcomes, though for different reasons. That is, in most cases couples will choose to implant the preembryos that yield the healthiest children, either because they are concerned impersonally with the *quality* of their future child's life or because they are concerned personally with the quality of *their* future child's life. The difference here, then, is primarily one of emphasis and motivation.

However, when we consider the couple who are considering implanting a preembryo that would result in a deaf child, Parfit might be able to provide somewhat stronger constraints than Heyd. While Parfit would not normally have moral grounds forcibly to keep a couple from making such a choice, he could plausibly make an argument that a deaf child would live a lower quality of life than a hearing child. He might also be able to constrain such choices based on their negative effects on the gene pool generally and on the implications of such choices for human survival in the distant future. Heyd, on the other hand, would be hard pressed to constrain this couple with a moral argument. He could argue that raising a deaf child might not be in their best interests, all things considered, but in this case the couple is already convinced that it *is* in their interests to have such a child. More plausibly, he might insist that the couple who intentionally makes such a choice should not be permitted to shift its extra child-rearing costs to the wider society.

Nevertheless, for the majority of couples presenting for genetic counseling, both of these positions would probably lead to similar outcomes (that is, if given the choice, we can safely assume that most couples will in fact *not* want to give birth to genetically affected children).

When we move to the collective level, however, problems develop that I believe might tip the theoretical "cost-benefit" balance slightly in favor of Parfit's impersonal approach. Suppose, as I am doing for this book, that we are concerned not only to evaluate a couple's use or misuse of a given genetic diagnostic or therapeutic technology relative to their future children, but that we are also concerned to evaluate the research policy that makes possible tens of thousands of such uses or misuses. Parfit's impersonal approach to value, especially as it is qualified by the requirement that agents adopt an agent-neutral perspective when considering such choices, will be able to guide such deliberations. These policy makers will ask themselves essentially the same question that the couple asked above: how will such a policy contribute to the overall quality and quantity of life in the world? This question will not eliminate the difficult empirical and predictive work Parfit would need to do to answer this question, though the qualifications we reviewed above could lessen these burdens.

In any case, such empirical and predictive work is no less necessary for Heyd's position if he is to determine who is potential and who is actual through some coordinated choice mechanism. Moreover, I argued above that his genethics is not only essentially agent-centric with respect to value, it is also agent-relative with respect to the perspective agents ought to adopt when justifying particular choices. Thus, these agents may have different moral aims, any of which could affect their determination of who is considered actual and who is considered potential in future generations. The determination of this question is very important for Heyd, because it will greatly influence his evaluation of allocative outcomes. Consider, for instance, the three-phase future I reviewed in chapter 3. If phase two generations are composed largely of actual or non-contingent persons from the perspective of the relevant agents, then he may evaluate HGP's research in very different terms than he would if those generations were composed largely of potential or contingent future persons. My concern, however, is that I do not believe he can answer this question; that is, he cannot tell the policy makers who is actual and who is potential because he cannot tell us which agent (that is, which agent's perspective) ought to be adopted when making the choice. On the other hand, all Parfit needs to know is that potential (contingent) persons are involved in a given choice relative to HGP's future phases; the question of who is actual and who is potential is irrelevant.

In the end, then, I find Heyd's position less persuasive than Parfit's. Genethics fails on the collective level, and with Parfit, I believe such a failure might be grounds for rejecting the position altogether. However, before I settle on this conclusion, I want to consider the theological implications of the discussion. For while I confess to being somewhat more persuaded by the impersonal approach to value, this creates a possible dilemma for me theologically. An impersonal approach to value, seen in theological terms, seems to imply that God serves an independent scheme of value rather than being viewed as the source or ground of value. For God to be seen as the source or ground of value requires, I will suggest, that we adopt the person-affecting approach to value. Fortunately, there may be a way out of this seeming dilemma. It may be possible to develop an approach to value that is formally person-affecting, substantively theocentric, and qualified in agent-neutral terms. Such an approach might permit a person-affecting approach to succeed on the collective level, though it too will have certain "costs" associated with it.

NOTES

1. Derek Parfit, *Reasons and Persons*, p. 444.

2. David Heyd, *Genethics: Moral Issues in the Creation of People*, p. 117.

3. In addition to the authors used below, I am indebted to Professor Jon Gunnemann's unpublished notes on consequentialism and public policy formation, and to R. M. Hare's "Public Policy in a Pluralist Society," in *Embryo Experimentation*, eds. Peter Singer, Helga Kuhse, Stephen Buckle, Karen Dawson, and Pascal Kasimba (Cambridge: Cambridge University Press, 1990), pp. 183-194.

4. This is not to suggest that the problem of contingent future persons does not challenge deontological theories of morality as well. However, David Heyd argues—persuasively, I believe—that deontological understandings of morality are, in principle, incapable of addressing the problem of contingent future persons. See, for example, his discussion of John Rawls and Kant in *Genethics: Moral Issues in the Creation of People*, pp. 41-55. This argument applies to theologically-based deontological ethics as well.

5. Parfit, Reasons and Persons, p. 24.

6. The traditional concerns of ethics—desires, dispositions, beliefs, emotions, in fact, *anything*—can be rendered relevant to the consequentialist insofar as they make outcomes better or worse. Parfit, *Reasons and Persons*, p. 25.

7. This understanding of consequentialism is derived from Frankena's discussion of teleological theories of morality. See William K. Frankena, *Ethics*, 2nd ed., Prentice-Hall Foundations in Philosophy Series (Englewood Cliffs, NJ: Prentice-Hall, Inc., 1973), pp. 14-15.

8. As a result, consequentialism makes "the deontic [that which an agent is morally obligated to do] dependent on the evaluative." See David McNaughton and Peirs Rawling, "Honoring and Promoting Values," *Ethics* 102 (July 1992): 842.

9. Adopted from Frankena, *Ethics*, pp. 14-15.

10. Other criteria are necessary to avoid a logical regression; at some point, we must make an argument about goods or value. This point is made by Bernard Williams in his "A Critique of Utilitarianism," in *Utilitarianism: For and Against*, J.J.C. Smart and Bernard Williams (Cambridge: Cambridge University Press, 1973), pp. 77-150.

11. John Rawls, *A Theory of Justice* (Cambridge, MA: The Belknap Press of Harvard University Press, 1971), pp. 22-27.

12. Robert N. Proctor, "Genomics and Eugenics: How Fair is the Comparison?," in *Gene Mapping: Using Law and Ethics as Guides*, eds. George J. Annas and Sherman Elias (New York: Oxford University Press, 1992), p. 81.

13. See above, pp. 56-57.

14. David Heyd, *Genethics: Moral Issues in the Creation of People* (Berkeley: University of California Press, 1992), p. 80.

15. Ibid. Emphasis added.

16. Ibid.

17. Ibid.

18. Ibid., p. 6.

19. Ibid., p. 3.

20. Derek Parfit, *Reasons and Persons* (Oxford: Oxford University Press, 1984), p. 370.

21. I am not trying to offer a defense of Parfit's consequentialism here. An adequate defense of consequentialist moral theory, if one were even possible, would take us too far afield.

22. See, for example, Daniel Holbrook, *Qualitative Utilitarianism* (Lanham, MD: University Press of America, 1988), p. 5.

23. Greatest expected goodness is equal to the sum of the values of each possible good effect multiplied by the probability that an act will produce it, minus the sum of the disvalues multiplied by the probability that an act will produce it. Parfit, *Reasons and Persons*, p. 25.

24. Parfit, pp. 30-33. Parfit's distinction between collective and individual consequentialism parallels what others call "rule" and "act" consequentialism.

25. Bernard Williams states the aim of any moral theory is to resolve conflict, in the sense that it provides some "compelling reason" to accept one intuition or argument over another. Bernard Williams, *Ethics and the Limits of Philosophy* (Cambridge, MA: Harvard University Press, 1985), p. 99. He constrasts this aim with an historical or a sociological approach to moral theory, which aim at *understanding* such conflict.

26. The empirical argument behind this case may be arguable, of course, but it illustrates the kinds of concerns Parfit has with individual forms of consequentialism.

27. Some moral philosophers believe agent-neutrality actually defines consequentialism as a moral theory, in constrast to agent-relative moralities which are viewed as deontological. See, for example, McNaughton and Rawling, "Honoring and Promoting Values."

28. Another way actions may be "impersonal" concerns how very small harms may be imperceptibly done by individuals that, taken together on the collective level, may be perceived to be very harmful. See Parfit's discussion of "The Harmless Torturers," *Reasons and Persons*, pp. 78-82.

29. Parfit's concern to argue in formal terms also contributes to the abstract quality of his discussion.

30. See Bonnie Steinbock, *Life Before Birth: The Moral and Legal Status of Embryos and Fetuses* (New York: Oxford University Press, 1992), p. 38.

31. Ibid., pp. 35-356.

32. Note that we do not need to know that the child will actually be harmed; it is enough that she *might* be. Ibid., p. 356. Time-neutrality also has implications for the evaluation of very small or imperceptible effects. See Parfit's discussion of "Five Mistakes in Moral Mathematics," pp. 67-86.

33. Ibid., p. 357.

34. See also his discussion of "Different Attitudes Toward Time," Parfit, *Reasons and Persons*, pp. 149-186.

35. Ibid., p. 351.

36. This claim holds, I believe, regardless of the substantive views we might have of persons. Again, this is not a claim that genetic endowment is sufficient to account for identity, but merely a claim that it is necessary.

37. Perhaps even for a night, but Parfit makes the weaker claim to make a stronger argument. Ibid., pp. 351-355.

38. Ibid., p. 359.

39. Ibid., pp. 361-364.

40. The claim that different policies lead to different populations of future people is an empirical claim and probably not one that most of us would dispute. What does prove to be problematic is the *moral relevance* of this claim. We will

see below that Heyd believes Parfit has over-emphasized its importance. But because of the ubiquitious effects of modern bureaucratic decisions, Parfit believes that many policies of modern states must be treated as different people policies. In Parfit's defense, his argument holds if a policy leads to populations with only a one-person difference between them.

41. This is a key distinction for Parfit that is determined primarily on the basis of imaginary self-reported preferences.

42. Ibid., p. 386.

43. Compare, however, Jan Narveson's defense of utilitarianism, in his "Utilitarianism and New Generations," *Mind* 76 (1967): 62-72. This article probably initated the contemporary debate on future generations and their contingency.

44. Arguments about "instrinsic" goods are problematic for some consequentialists, but Parfit finds them acceptable.

45. Ibid., p. 390.

46. Ibid., p. 410.

47. See Ibid., "Appendix G," p. 487.

48. Ibid., p. 412.

49. Ibid., p. 420.

50. Heyd, *Genethics: Moral Issues in the Creation of People*, pp. 55-64.

51. Heyd, *Genethics: Moral Issues in the Creation of People*, p. xi. Emphasis added.

52. Heyd is refreshingly explicit about his method and its limitations. Ibid., pp. 70-80.

53. Ibid., p. 82.

54. Ibid., pp. 97-98. This is not to say that he defends his position on the basis of a definition alone. The rest of the argument tries to build a plausible case for making this sort of distinction.

55. Ibid., p. 99.

56. Ibid., p. 97.

57. Ibid., p. 99.

58. Ibid., p. 98.

59. Ibid., p. xii. Emphasis added. Genethics is understood by Heyd in a much broader sense than the authors of a book by the same title understand their term. Compare D. Suzuki and P. Knudtson, *Genethics* (Cambridge, MA: Harvard University Press, 1989), where "genethics" refers to problems of modern genetics. Heyd's term encompasses this and much more, and in my estimation his book offers a much more suggestive way to address genetic problems.

60. Heyd, *Genethics: Moral Issues in the Creation of People*, p. xii.

61. Ibid., pp. 4-5. His views on the creation story are unorthodox but Heyd is unconcerned; he is using the story metaphorically, not theologically.

62. Ibid., p. 6.

63. Ibid., pp. 4-8.

64. Ibid., pp. 84-87.

65. Ibid., p. 79.

66. Ibid., pp. 103-106. This difference between Heyd and Parfit raises a question about the empirical scope of the problem of contingent future persons that I do not pursue; however, it is one worth considering at some point in the future.

67. Ibid., p. 98.

68. Ibid.

69. Ibid., p. 124.

70. Ibid., pp. 106-115.

71. By "convoluted" I do not mean that Heyd's arguments are unclear; indeed, they are remarkably lucid. Rather, I mean that it becomes impossible to determine who the relevant agent ought to be on the collective level.

72. Recall, however, that Parfit used the discussion to test the *theoretical* limits of his impersonal approach to value.

73. Ibid., p 154.

74. Ibid., pp. 127-159.

75. Ibid. p. 131.

76. The feminist author Iris Marion Young raises similar types of questions in the first chapter of *Justice and the Politics of Difference*.

5

Personal and Impersonal Theocentric Approaches to Value

> If concrete actions promote a value, they are rescribable. If they generally attack a value, they are generally proscribed. If they always attack a value, they are always proscribed. Now, whether an action attacks a value or not, whether it is loving or not, is determined by its relation to the order of persons.[1]]
>
> Richard A. McCormick

> The theocentric construal of reality...presses us to expand the scope of the wholes that are taken into account in any normative proposal for human action. Such enlargement greatly complicates ethics...[2]
>
> James M. Gustafson

In the previous chapter we saw that the problem of contingent future persons forces us to choose between two fundamental and competing conceptions of value, what are called impersonal and person-affecting or person-oriented approaches to value. While both of these conceptions are essentially formal, we saw that Heyd developed his genethics in anthropocentric terms. He acknowledges, however, that a person-affecting approach to value could also be theocentric. In this chapter, I will attempt to explore and develop this theocentric option.[3]

I will argue that theological ethics or moral theology, *as they are typically construed* in the Christian tradition, cannot adequately account for contingent future persons. This is not, however, because Christian theological ethics are typically construed in person-affecting terms with respect to an approach to value, but because they are typically construed in agent-relative terms with respect to the perspective agents ought to adopt when making moral judgments. Recall that it is not Heyd's person-affecting approach to value that rendered his approach self-defeating, but the apparent failure on the collective level of his agent-relative genethics to tell us who is potential and who is actual. The same problem arises for traditional theological ethics. It is possible, however, to conceive of a theological approach to the problem of contingent future persons that is person-affecting with respect to the nature of value but agent-neutral with respect to the perspective agents ought to adopt when making moral judgments. I call this position an "impersonal theocentric" approach to value, and contrast it to the traditional approach, which I characterize as a "personal theocentric" approach. In order to develop these distinctions, I utilize three authors, Richard A. McCormick, William P. George, and James M. Gustafson, with McCormick and George representing the personal theocentric approach and Gustafson the impersonal theocentric approach.[4]

Unfortunately, as in the previous chapter, the development of these two theocentric approaches to value is complicated by problems of terminology. I refer particularly to the way Gustafson and others have criticized the Christian tradition for being overly "anthropocentric." I accept the point of the critique, namely, that the tradition unjustifiably views humans as being central to God's purposes for the world, but I also believe it is necessary to clarify the use of the term anthropocentric in this context. Gustafson contrasts his "ethics from a theocentric perspective" to the so-called "anthropocentrism" of current secular or humanistic ethics and of traditional theological ethics. However, I believe the contrast is more accurately applied only to the former. With respect to fundamental conceptions of value, the Christian tradition is theocentric in exactly the same way Gustafson's approach is theocentric; it views God as the source or ground of value, and thus, to the extent that the tradition gives humans a central role to play in God's purposes, it does so for *theocentric* reasons.

The implication of this clarification is that the Christian tradition cannot be viewed as anthropocentric in the same sense that Heyd's humanistic approach to value is anthropocentric. The tradition does not hold that value is dependent on the existence of *human* valuers, as Heyd does; rather, it holds that value is dependent only on the existence of *God*, and that the central role accorded humans by the tradition is warranted by reference to God. Again, it is this last point, the centrality of humans for God's purposes, that in my view Gustafson rightly contests. I am merely suggesting here that the central role traditionally accorded to humans is perhaps better distinguished by reference to the perspective agents ought to adopt when making value judgments, and that Gustafson's use of the term anthropocentric to describe the tradition is somewhat misleading on this point.

In any case, once these issues are clarified, I believe Gustafson's understanding of God, and the implications of his understanding for theological ethics, can in fact be distinguished from the tradition in ways that permit us to address contingent future persons directly. The distinguishing characteristics of his position are both substantive and methodological. Substantively, Gustafson views God in more impersonal terms, and this gives him grounds both for admitting more contingency into his theological ethics and for de-centering humans in his understanding of God's purposes for the world. Methodologically, I will argue that his ethic is agent-neutral, and more concerned with consequences or states of affairs than is often the case in the tradition, and that these aspects of his method will permit us to incorporate many of Parfit's impersonal qualifications. Taken together, then, these features of Gustafson's theocentric ethics will be helpful in addressing the problem of contingent future persons with what remains formally a person-affecting approach to value. We begin by investigating in what sense(s) a theocentric approach to value of any type, personal or impersonal, can provide ethical guidance for human agents. For this, we revisit Heyd's discussion of the first Genesis myth.

GENESIS REVISITED

David Heyd claims that the two fundamental theories of value we explored in the previous chapter are mutually exclusive and exhaust the formal possibilities open to us. I do not believe this claim can be debated. We must, therefore, decide between them if we are concerned to develop a consistent ethic. Heyd, we saw, chooses a person-affecting approach to value, and is thus forced to distinguish between actual and potential persons. Then, in order to map the implications of this approach for genethics, Heyd distinguishes between a pure and an impure type of genesis choice (any choice that involves the "creation of people"). Pure genesis choices are illustrated by God's choices at the original creation.

They are utterly agent-centric and thus utterly unconstrained, God (the agent in this case) being the only actual person who can be affected by the choice. Impure genesis choices are primarily but not exclusively agent-centric; they may be constrained empirically by various blocks of interests and morally by the probable effects of the choice on other actual persons.[5]

Here I am primarily interested in recalling how Heyd moves from his pure or ideal type of genesis choice to his impure type. He makes this move on the basis of an analogy between God's creation of the first humans (as moral subjects, like God) and our creation of subsequent humans. This analogy is important for Heyd, for on the basis of it he develops a person-affecting approach to value that is substantively and exclusively anthropocentric. Heyd believes we live in a world where God cannot plausibly serve as a basis for a person-affecting approach to value; humans are thus the only persons who can serve in this capacity.

> [I]f God's existence is denied, human beings are left as the only reference point or "source" of value in the world. They themselves become God, in the sense that they have full sovereignty over the existence of value.[6]

As a "concise formulation" of his so-called axiological volitionalism, Heyd quotes Protagoras: "Man is the measure of all things."[7]

However, in opting for a person-affecting approach to value that is substantively anthropocentric, Heyd acknowledges that the term "person" logically need not be limited to human persons. That is, a person-affecting approach to value need not be anthropocentric, or not exclusively so. It can also be theocentric.[8] By definition, a theocentric approach to value holds that God's existence is a necessary condition for the existence of value. It thus finds its center, its source, and its primary reference point in God. For those who remain persuaded that belief in God can provide a basis for an approach to value, and thus ethics, this possibility suggests that a theocentric "genethics" might be developed in ways that Heyd does not elaborate.[9] He does, however, make one important theological observation that deserves to be reiterated.

Heyd observes that a theocentric approach to value preserves the important theological claim that *God constitutes value rather than serves it.*[10] We can debate the meaning of the term "person" as it is applied (analogically) to God, but I do not believe we can claim that God is both the source or ground of all value and at the same time claim that God serves an independent scheme of value. However, the latter claim is implied by an impersonal approach to value. This suggests then that a theocentric approach to value cannot be impersonal with respect to value in the sense advocated by Parfit; rather, it must be person-affecting, as Heyd in fact argues. This said, we must still ask how a theocentric approach to value can serve as the basis of an ethic for human agents. Heyd is helpful here, too.

Recall that Heyd is careful to distinguish between the *nature of value*, that is, the conditions of value attribution or value as a relational property, and the *perspective* from which value judgments and attributions are made, that is, the way value judgments and attributions are grounded or validated.[11] His person-affecting thesis pertains only to the former category, that is, to the nature of value; it claims that the existence of persons (human or divine) is a condition of value attribution (which is, of course, why potential persons are considered outside the moral domain). A value judgment, however, can be made from the perspective of any relevant agent. Thus, we must ask how theological ethicists treat this question of perspective when either God or a particular (human) individual (or group of individuals) could conceivably be the agent(s) in question.

Theological ethicists are required to specify more carefully *whose* perspective is to count in making value judgments, the divine or the human. This task is complicated by the fact that most contemporary theologians want to base their value judgments on claims about God and God's purposes for the world without at the same time claiming that they view the world from God's perspective (a claim which would be seen as religious hubris and as epistemologically naive). Thus, theologians must make a further distinction between the *basis* on which something is valued and the *perspective* from which it is valued.[12] In this context, the term theocentric refers to the basis on which something is valued; that is, whatever is valued is valued on the basis of a belief that God values it. However, because a theocentric approach to value is intended to serve as the basis of an ethics for *human* agents, whatever is valued must also account for the perspective of the human agent or agents. Humans gain knowledge of God's purposes for the world through an interpretation of "revelation" or of a variety of other so-called "natural" sources. In this way, then, human agents can make judgments that are based in their beliefs about what God values without being required to make the implausible claim that they view the world from God's perspective.[13]

Now, with these distinctions in place, we can return to Heyd's analogy between divine and human agency. We can accept with Heyd that the creation of the *first* humans could be good only for God; however, we need not assume that the creation of *subsequent* humans by human agents need be considered good only for humans (that is, good only in anthropocentric terms). Persons created subsequent to the first creation can (also) be valued on the basis of a reference to God (that is, good in theocentric terms). Further, we can accept with Heyd that impure genesis choices are analogous to God's pure genesis choice in the sense that humans now possess *de facto* god-like powers over contingent future persons; again, however, we can do so without being forced to conclude that such choices have only human value referents that can serve *de jure* to constrain the use of these powers. Said differently, humans have obligations *to God* that can be figured into their genesis choices.

However, as I mentioned above, when a theocentric approach to value is used as the basis of an ethics, it is complicated by the fact that theologians hold substantively different views on the centrality and importance of human persons for God's purposes. While these different views each claim theocentric warrants, they also influence theologians' understanding of the perspective human agents ought to adopt making particular value judgments. Moreover, this choice about possible perspectives has implications for addressing the problem of contingent future persons; to uncover them I turn to the work of McCormick, George, and Gustafson.

A PERSONAL THEOCENTRIC APPROACH TO VALUE

Richard McCormick's fundamental criterion for judging the rightness of any action or policy is unequivocally person-affecting or person-oriented. I quote him in the epigraph of this chapter. He states, "...whether an action attacks a value or not, whether it is loving or not, is determined by its relation to the order of persons."[14] The "persons" he refers to are of course human, and this is a commitment he shares with the wider

Christian tradition. But, by giving humans such a central role in his moral theology, McCormick leaves himself open to criticism by theologians like Gustafson for being too "anthropocentric" (in the above-mentioned qualified sense). Gustafson believes that the tradition's so-called anthropocentrism limits the moral domain in unacceptable ways. I find McCormick useful here because his understanding of the tradition permits him to conceive of the moral domain in increasingly broader terms. He is thus a strong representative of the best the tradition can offer on this point, over against Gustafson's critique. Nevertheless, even McCormick's understanding of the moral domain may not be drawn widely enough to address the problem of contingent future persons. He thus provides a good "foil" for introducing Gustafson's impersonal approach, which permits a moral domain that is even more expansive than McCormick's.

THE CENTRALITY OF THE HUMAN PERSON

As a Roman Catholic moral theologian, the proximate warrant for McCormick's commitment to persons comes from the Vatican II document, *Gaudium et Spes*, which states,

> The moral aspect of any procedure [that is, any human act]...must be determined by objective standards which are based on the nature of the person and the person's acts.[15]

This is at root a "natural law" argument, but one that under the impetus provided by Vatican II has been historicized and considerably personalized in order to emphasize the natural or innate "dignity of persons."[16] This emphasis on personal dignity has created space for a variety of interpretations of the Second Vatican Council's phrase, the "nature of the person."

For his interpretation, McCormick follows Louis Janssens, who believes the "nature of the person" to be better specified by the "human person integrally and adequately considered."[17] Janssens further amplifies his own terminology to mean the human person in his or her eight (8) "essential aspects." Considered singularly, the human person is:

> a subject (normally called to consciousness, to act according to conscience, in freedom and in a responsible way); [and,]
>
> a subject embodied.

Considered relationally, persons are:

> ...embodied subject[s] that...[are]...part of the material world;
>
> ...essentially directed to one another (only in relation to a Thou do we become I);
>
> ...[created with a] need to live in social groups, with structures and institutions worthy of persons;
>
> ...called to know and worship God;
>
> ...historical being[s], with successive life stages and continuing possibilities; [and,]
>
> ...utterly original but fundamentally equal.[18]

This said, the ultimate warrant for McCormick's commitment to persons is finally theocentric. That is, it is authorized by beliefs about God and God's purposes for humans, and supported through an interpretation of the natural law. Thus, for McCormick it is finally, though not exclusively, God who authorizes an approach to value that regards human persons as so foundational to ethics. He states:

> [T]he morality of our conduct is determined by the adequacy of our openness to [values that define human well-being or flourishing], for each of these values has its self-evident appeal as a participation in the unconditioned Good we call God. The realization of these values in intersubjective life is the only adequate way to love and attain God.[19]

For my purposes, McCormick's position regarding the centrality of persons is useful for how it illustrates one way of grounding what I call "personal theocentrism." I use this phrase to capture what I take to be a general approach to persons in the Christian tradition, with its sweeping soteriological drama of creation, fall, restoration, and consummation. Personal theocentrism makes at least three claims. It claims: that God's purposes for creation are benevolent (in the sense that God establishes conditions for the well-being of creation and in contrast to a malevolent or utterly indifferent God); that humans enjoy a unique status as moral subjects or persons, in contrast to other living species on Earth; and, key to this position, that God's purposes for creation are ultimately coincident with human good.[20] It is this third claim, that God's purposes for creation are ultimately coincident with human good, that gives this position its distinctiveness over against the "impersonal theocentrism" that I develop below using Gustafson's work. It is the basis for Gustafson's claim that the tradition is "anthropocentric." He says:

> The dominant strand of piety and theology...has focused on the grandeur of man, on the purposes of the Deity for man, and primarily on the salvation and well-being of man.[21]

I believe Gustafson's claim about the tradition is generally true, and yet I wish to emphasize, in contrast to some of his rhetoric, that this anthropocentrism can be interpreted in ways that permit the moral domain to be drawn with very broad boundaries. We can see how this breadth is achieved in one stream of the tradition by briefly examining McCormick's methodology.

Proportionalism

McCormick's method has been given the name "proportionalism" by its detractors, and he has adopted this designation himself in his more recent writings.[22] Proportionalism is superficially similar to consequentialism, and often confused with it, though McCormick sees consequentialism (at least, in its less qualified forms) as more reductionistic than proportionalism. In any case, proportionalism is teleological in structure and is perhaps best illustrated under conditions of moral ambiguity.

Because human finitude and sin impose limitations on agents and on their ability to effect desired outcomes, moral choices are often ambiguous in the sense that many actions lead unavoidably to a mix of values and disvalues. HGP is a good example of this mix, as we saw in chapters 2 and 3. When such choices arise, McCormick argues that the key question is whether there is a *proportionate* reason to perform an activity which issues simultaneously in such mixed results. By this, McCormick is asking whether there is enough value to justify the resulting and unavoidable disvalues. Such a question implies that simply causing (ontic, non-moral) evil through an action does not necessarily make this action (morally) wrong. Thus, for example, the fact that we can predict some harms from HGP does not necessarily imply we ought not to pursue it. The point here, and what makes this approach so objectionable to McCormick's conservative Roman Catholic critics, is that his view of what makes an action right or wrong makes it very difficult, if not impossible, to declare an action to be absolutely prohibited.[23] What makes an action right or wrong, all things considered, is whether there is or is not a proportionate reason justifying it.[24] Such a reason can be discerned only by examining an "action as a whole" or in its "totality."

The term totality is a technical term in Roman Catholic moral theology. It is used to describe the relevant moral domain, and it illustrates how a personal theocentric approach to value need not define the moral domain in narrow terms. Consistent with the relational understanding of human persons developed by Janssens, personal acts are given an expansive definition. An "action as a whole" includes the actual physical act in question, the agent's intention, the situation or circumstances, and the act's consequences.[25] Moreover, the consequences of the act can extend indefinitely into the future, and the circumstances or conditions that give rise to and influence the act are defined so as to include an understanding of the individual as part of a larger natural and social community. The individualism that in our culture is so often a correlate of an approach to value that makes persons so central can be constrained as well. When an action that benefits individuals involves risk of harm to basic social institutions or collectivities, McCormick argues that the institution should generally take precedence over the individual.[26] He believes in such cases that collective personal goods (institutions) generally ought to be ranked over individual personal goods, as these collective goods provide the conditions for the realization of individual goods.

This broad ranking of goods is further qualified by strong prudential elements, especially when McCormick attempts to evaluate public policy in pluralistic societies. Thus, in addition to the moral criteria surrounding the centrality of persons and the importance he ascribes to the collective bodies that support persons, he proposes the criterion of "feasibility" when considering public policies[27] and a reliance on "procedures" where no consensus exists concerning the implications of novel policy efforts.[28] My point, again, is that a traditional understanding of the centrality of persons for God's purposes can support very expansive understandings of the moral domain. I now want to ask is whether the moral domain drawn by McCormick's personal theocentrism can adequately address the problem of contingent future persons.

THE MORAL STATUS OF THE PREEMBRYO

In asking this question, I am especially concerned with McCormick's endorsement of Janssens' claims that persons are "embodied subjects" and "utterly original." While these claims help ground strong protections for nascent human life in the Roman Catholic tradition, they also permit McCormick, as a Roman Catholic moral theologian, to step onto a "slippery slope" regarding the status of preembryonic human life. McCormick's discussion of the preembryo occurs in the context of considering the moral licitness of *in vitro* fertilization (IVF) and genetic therapies. In both cases the moral status of the preembryo is held to be key to his conclusions, in that its status largely determines what morally can be done to it.[29]

As a procedure, IVF is prohibited by the Roman Catholic Church on grounds that the preembryo is produced in an non-natural act—a so-called "physicalist" interpretation of the natural law. Moreover, the Church prohibits experimental manipulations on all forms of nascent human life, except in those rare instances where the action might be construed as directly therapeutic and one of last resort for a particular embryo.[30] The basis for this position is the belief that personhood should be attributed to human life "from the moment of conception."[31] Thus, in the Church's official view, neither the *in vitro* preembryo nor the *in vivo* embryo should be subjected to manipulations that threaten its existence, and it should not be used as a means to any end other than one that promotes its own good.

Now, McCormick is clearly disposed toward arguments that protect early human life. However, his less "physicalist," more personalistic view of the natural law permits him to view the *in vitro* production of preembryos as a potentially licit procedure, at least for married couples who are trying to overcome infertility. Rather than an artificial barrier to the "natural" love expressed in the sexual conception of children, it can be seen as an extension of it. "Not everything," he says, "that is artificial is unnatural."[32] Also, unlike his more conservative colleagues, it is not clear to McCormick that preembryos ought to be regarded as persons in the fullest sense of the term, with a full complement of rights. It is this view that is of particular interest to me.

On the question of the personhood of the preembryo, McCormick challenges the notion that individuality can be meaningfully ascribed to preembryos just at or after conception, whether considered *in vivo* or *in vitro*. While McCormick grants that such evaluations can never be made on scientific grounds alone, he is nevertheless concerned that the Roman Catholic Church's absolute prohibitions against the production of and experimentation with preembryos are based solely on a "genetic" notion of individuality. These notions are not plausible, McCormick believes, in light of current "developmental" notions coming from more recent biological research on human reproduction. In developmental terms, individuation takes place not at conception but after implantation, when there is no further possibility of twinning (usually after the fourteenth day). Thus, he suggests, in response to reasons given for prohibiting IVF by the Congregation for the Doctrine of the Faith (CDF), the proper answer to its rhetorical question, "How can a human individual not be a human person?," is "by not being—yet—a human individual."[33] Thus, if persons are understood as "embodied subjects" and as "utterly original" (which I take to mean "unreplaceable"), and if originally cannot be guaranteed developmentally until after the possibility of twinning is past, then strict personhood cannot meaningfully be ascribed to preembryonic human life.[34]

Before I pursue the implications of this argument for addressing the problem of contingent future persons, I want to review McCormick's treatment of germ-line gene therapy. Unfortunately, he has little to say about it. With respect to genetic therapy in general, he can (predictably) accept the licitness of somatic cell gene therapy when safety, efficacy, cost, and access issues are accounted for, but raises what are merely rhetorical questions about germ-line gene therapy. His concerns are primarily centered on the unpredictable effects of germ-line gene therapy on future generations.[35] These unpredictable effects presumably serve as reasons to prohibit such research altogether or to move forward with it very cautiously (in light of his previous statements about medical research, the latter is probably more plausible).[36] However, he does not develop these thoughts. Instead, he observes that this therapy is not now technically feasible and does not discuss it further.

McCormick's brief comments and rhetorical questions do, however, suggest that future generations in general are important to him and that they can be meaningfully treated under a personal theocentric approach to value. With respect to evaluating genetic therapies in general, McCormick reiterates his commitment to persons.

> The central question is always this: Will this or that intervention (or omission, exception, policy, law) promote or undermine human persons "integrally and adequately considered"?[37]

However, he continues,

> The answer to such a question...cannot be deduced from a metaphysical blueprint of the human persons. It is necessarily inductive, involving experience and reflection upon it.[38]

This statement follows from his proportionalism (and, again, it contrasts with the deontological prohibitions of his conservative colleagues). However, our experience with germ-line gene therapy is obviously limited, especially in humans. Thus, because of our limited experience and because we lack societal consensus about such questions, McCormick argues that we can do no more than resort to agreed upon procedures for making decisions, though he does not recommend a mechanism of coordinated choice as Heyd does. Rather, he suggests that we continue studying the question of genetic therapies, apparently along the lines now under consideration in Congress.[39]

> The matter is too important to be left to the ordinary political dynamic...[S]ome further *public* mechanism of ongoing deliberation [is necessary]...because it is a common human future we are deliberating.[40]

In the end, then, with respect to the moral status of the preembryo, McCormick offers a cautious and somewhat ambiguous statement to the effect that he does not believe "nascent life at this [pre-individuated] stage makes the same demands for protection that it does later."[41] We might infer from this discussion, however, that he would grant some protections to preembryonic life on the basis of the fact that, under the proper conditions, preembryos will develop into persons in their own right. Thus, for example, he hints that it might be possible to justify treating the preembryo *as if* it were a person, but does not develop this thought.[42]

Nevertheless, once McCormick opens the possibility of the licit production of preembryos, he must confront the question of what criteria should be used in selecting preembryos for transfer. He is aware that with IVF often not all preembryos are transferred, and he is aware that many that are transferred will die. If all cannot or should not be transferred, can personal theocentrism provide guidance to individual or collective agents for selecting among preembryos or for altering the germ-lines of affected preembryos? Or, if we believe that it might be wrong for couples deliberately to create and transfer genetically affected preembryos, is there any way McCormick can consistently develop constraints on such an act? Or, assuming Roman Catholic couples remain opposed to aborting affected fetuses, might they resort to germ-line gene therapy to alleviate the suffering of their future (contingent) child? If they did, on what moral basis could they decide such questions, knowing that they cannot meaningfully be said to harm the child they bring into existence? To the best of my knowledge, McCormick does not speculate on questions like these. If he did, however, I believe the best he could offer, given what I am calling a personal theocentric approach to value, would be indirect constraints similar to those developed by Heyd.

Consider, on the individual level, the case from chapter 1 of the couple who is trying to decide between implanting an affected preembryo that, if it survives, will result in a deaf child and an unaffected preembryo that will result in a healthy child. If the preembryos in question are regarded as actual or non-contingent persons (as official Roman Catholic teaching holds), both would claim equal protection. But, if they are both regarded as potential or contingent, as McCormick's position seems to imply, it is not clear how a personal theocentric approach to value could guide the couple's choice. If I am correct, they would need to have some reason to believe that God values one particular future person over another, since in such a choice both would not be born (and could not be born, if a germ-line intervention on one preembryo were being contemplated in order to achieve the same end). I cannot see how such a belief could be grounded in personal theocentric terms. We could, of course, follow Heyd's suggestion and consider the probable effects of the existence or non-existence of the future child on the couple and on other persons. But this consideration does not require a theocentric approach to value. In short, compared to Heyd's anthropocentric genethics, personal theocentrism seems to provide no additional (theocentric) reasons to guide individual agents in such situations. This fact may not worry McCormick, since he holds that religious beliefs are morally influential primarily on the level of motivation and do not normally figure in particular choices. And yet, I suspect that many theological ethicists and moral theologians will resist the arbitrariness of such choices (as seen from a personal theocentric perspective) or the inherent egoism of these choices (as seen from Heyd's anthropocentric perspective).

Interestingly, however, a personal theocentric approach may provide some guidance to agents when we move to the collective level. This possibility is suggested by William P. George's work, the one moral theologian I have discovered who has commented on the problem of contingent future persons. George makes a brief reference to contingent future persons as he explores the thought of Thomas Aquinas for theological resources with which to address our contemporary concerns for future generations. He argues that Aquinas can be interpreted so as to support a general claim that we ought to care for future generations, and I find him convincing on this point. However, George also seems to assert that *contingent* future persons ought to receive such care as well, simply *because* their existence is contingent and they are thus vulnerable to our acts toward them.

> Future persons claim attention precisely because present agents have a real say—albeit partial—in whether, or how, future persons exist.[43]

George claims this regard is a demand of prudence for Aquinas, and that it is grounded in charity. I am not competent to judge his claim regarding Aquinas' position in this case; we might ask, for instance, if prudence demands that contingent future persons be treated "as if" their existence were non-contingent. Would this help us select between them on the individual level? Not in any way that I can discern. It may be prudential to care for contingent future persons in the sense that this care may contribute to the formation of agents who possess characters or virtues that are disposed toward protecting human life generally, but it is not clear what difference such good characters or virtues would make in justifying particular choices between particular preembryos, or in justifying particular choices regarding therapeutic interventions on the genetic level. Thus, though I believe he is correct to assert our responsibility for future persons, George may underrate the force of the problem of contingent future persons.

This problem notwithstanding, George may in fact offer a way to constrain agents on the collective level. He claims that we must link our general concern for future generations to our obligations to care for creation generally. He claims that:

> Prudence, grounded in charity urges regard for future human beings *along with* their environment as these are related to God.[44]

This move is certainly consistent with a personal theocentric approach to value. Creation is viewed as basically "good" in the personal theocentric approach to value. That is, it was created good "for God" and, with the creation of humans as subjects "like God," God permitted it to be good "for humans" as well. However, in this approach to value humans occupy a central role in God's purposes for creation, and thus the other goods and species of creation are viewed as *instrumental*[45] to human good, and this not simply in a *de facto* sense due to human dominance but in a *de jure* sense due to God's command and blessing.[46] This has implications for the human stewardship of creation and for addressing the problem of contingent future persons on the collective level.

In a personal theocentric approach to value, the motivation for acting as stewards of creation is primarily theocentric. It is an obligation humans owe to God because God is believed to value creation. Thus, while creation is valued instrumentally, we would not be forced with Heyd to claim that the world is utterly valueless unless human agents invest it with value; presumably, a world without humans is still valuable to God. However, in this approach to value, God is also believed to authorize the stewardship of creation primarily *for the sake of humans* (that is, actual or non-contingent humans). Thus, whoever is born in the future might find themselves enjoying the fruits of a creation that has been adequately conserved and protected.[47] Again, this suggests that a personal theocentric approach to value can support a very broad interpretation of the moral domain; it need not necessarily imply the exploitation of creation in the sense that Gustafson and other critics sometimes suggest.

Now, however, we must ask whether such a prudential linkage of care for future persons to the stewardship of creation will provide reasons on the collective level for agents to choose what Parfit called the conservation policy over the depletion policy. I believe it might, but in a carefully qualified sense. By linking moral obligations to present and future humans with prudential concerns for creation generally, creation can be valued by present humans for future (actual or non-contingent) humans as an obligation owed to God; moreover, presumably God values creation impartially for whoever is born in the future.[48] On this basis, then, collective agents could be constrained in their choice of policies that seriously deplete natural resources for future generations, and future persons born as a result of this policy might, other things being equal, find their lives less diminished. But note again that the protection provided by this linkage would be aimed at actual or non-contingent future persons. Contingent future persons as such would not be included in the moral domain and could thus be accorded only indirect protection.[49] Again, the primary difference between this approach and Heyd's is motivational; creation is protected as an obligation owed principally to God rather than as an obligation owed to future persons. This implies that if the choice in question were redefined as a choice between different populations of contingent future persons (as opposed to being a choice between policies that conserve or deplete natural resources), a personal theocentric approach would pro vide no additional guidance, relative to Heyd's anthropocentric approach, in choosing one population of future persons over another. We could follow Heyd's lead and try to predict the effects of our choice between contingent future populations on actual or non-contingent future persons, but we could not escape the problem on which Heyd's approach stumbled: identifying who is potential and who is actual, and whose perspective ought to count in making this determination.

In sum, then, a personal theocentric approach to value clearly provides agents with a different motivation for performing moral acts that affect future persons, and it can cut against the instrumentalization of creation on the collective level by linking obligations to future people to a prudential concern for the creation on which these future people will depend. Nevertheless, it seems to provide no more guidance when choosing between particular contingent future populations than Heyd's anthropocentric approach to value. I now want to explore James Gustafson's less traditional approach to value, which I refer to as "impersonal theocentrism." This latter approach remains fundamentally person-affecting, but I believe it is qualified in ways that render it better able to address the problem of contingent future persons.

AN IMPERSONAL THEOCENTRIC APPROACH TO VALUE

For Gustafson, theology is an interpretation of, or a second-order reflection on, the world and the religious dimensions of human experience. Analogously, ethics is viewed fundamentally as reflection on the moral dimensions of human experience. This view of ethics is applied by Gustafson to both theological and philosophical approaches to ethics. Theological ethics are distinguished from philosophical ethics by how God is referenced as a way of understanding the world and the meaning of moral experience. Further, Gustafson claims that to be coherent theological ethics should be organized around a central theme, and to be comprehensive it should address four "base points." Gustafson's central theme is found in what he calls the "theocentric perspective," which I will discuss momentarily. His four base points include:

> the interpretation of God and God's relation to the world and particularly to human beings, and the interpretation of God's purposes;
>
> the interpretation of the meaning or significance of human experience—of historical life of the human community, of events

and circumstances in which persons and collectivities act, and of nature and man's participation in it;

the interpretation of persons and collectivities as moral agents, and of their acts; and

the interpretation of how persons and collectivities ought to make moral choices and ought to judge their own acts, those of others, and states of affairs in the world.[50]

In addition to these four "base points," which any theological ethic should address if it wants to claim comprehensiveness, Gustafson claims that Christian theological ethics can be further differentiated by how they use and understand four "sources." These sources include:

the Bible and the Christian tradition;

philosophical methods and principles;

the use of scientific information and other sources of knowledge of the world; and

human experience broadly conceived.[51]

His efforts at comprehensiveness notwithstanding, I believe Gustafson's distinctiveness as a theological ethicist is found especially in his treatment of his first, second, and fourth base points and the third source; that is, I believe his distinctiveness is found principally in his interpretation of God's rather impersonal relation to the world, his judgment that the meaning of human existence is inadequately captured by the tradition (his criticism of the tradition's so-called anthropocentrism), the weight he gives to consequences or states of affairs when making moral choices, and his reliance on scientific sources to interpret the purposes of God for creation. I am interested here in how these distinctive aspects of his approach permit him to argue for a view of the moral domain that is even more expansive than McCormick's and George's.

De-Centering the Human Person

Gustafson's project is both critical and constructive. He is critical of what he calls the anthropocentrism of contemporary philosophical and theological ethics. His constructive efforts are revealed as he develops a set of complexly qualified theocentric constraints on this anthropocentrism. Heyd's position is a good example of the type of philosophical or humanistic anthropocentrism that Gustafson criticizes (which is not to suggest he discusses Heyd's work as such), while McCormick's and George's positions are good examples of theological anthropocentrism. The latter, however, is complicated by its theological grounds, and I suggested above that it might be better represented by a term such as personal theocentrism. Personal theocentrism is distinguished by its claim that humans occupy a central role in God's purposes for the world and, particularly, that these purposes ultimately coincide with human good. Gustafson grants that the tradition's anthropocentrism has achieved much good, but he nevertheless believes it must be limited.

> Anthropocentrism has sustained the dignity and the rights of humans, collectively and individually. This cannot be denied. But even from a view that gives highest value to the human species, limits must be recognized.[52]

These limits or constraints are developed on a number of grounds.

First, working generally from his roots in the Reformed tradition, Gustafson sees the claim that God's purposes for the world coincide with human good as theologically presumptuous. Moreover, he believes it too easily permits religious faith to be turned into a device that merely satisfies human desires and addresses human anxieties. Such views unjustifiably push God to the periphery of faith and life. He states:

> My contention is that in our own time, as much as in any other, religion is propagated for its utility value to individuals and communities. Religious belief, trust, and practice, are offered as useful instruments for getting on well in the business of living, for resolving those dilemmas that tear individuals and communities apart, and for sustaining moral causes, whether they

> be to the right, the left, or the middle. Both individual pieties and social pieties become instrumental not in gratitude to God, the honor of God, or service of God, but to sustaining purposes to which the Deity is incidental, if not something of an encumbrance.[53]

Second, Gustafson is deeply influenced by certain scientific views that, in his estimation, seriously undermine the plausibility of the traditional claim that humans hold a central role in God's purposes or that God's purposes ultimately coincide with human good. For example, on the basis of his reading of contemporary biology and sociology, Gustafson believes that the traditional view fails to account adequately for the interdependence of Earth's biological and social systems. Thus, for instance, in criticizing how the first Genesis myth is typically interpreted in the tradition, he states:

> Interestingly, few theologians have read the passage to be an account of the dependence of human life on light and water, on seeds and plants, on animal life, all of which God saw was good (and not, I take it, good only for man).[54]

The interdependence of living and natural systems is a recurring theme in Gustafson's work, and he believes that the failure to account for it adequately has contributed dramatically to our reluctance to acknowledge constraints on the exploitation of nature and of other species for human ends. Essentially, Gustafson claims, this view of God's purposes for humanity is an attempt to fit a Ptolemaic view of religion into a Copernican universe. "Theology and religion," he says, "remain 'Ptolemaic,'" and for this reason remain less plausible intellectually and less defensible morally.[55]

Finally, while theology is always prior to ethics for Gustafson, this traditional theologically-based anthropocentrism is also troublesome to him for some of its ethical consequences. He believes that the ethics that follow from it are too easily corrupted into short-sighted self-interest or, worse, outright egocentrism.[56] In Gustafson's terms, anthropocentrism does not adequately account for the "whole" (his term for the moral domain), focusing instead too much attention on the importance of one "part," the human, and permitting this one part to view the other parts as instrumental to its good and advancement.

This said, Gustafson does not imagine that the human perspective can or should be escaped altogether, but neither does he claim that his "ethics from a theocentric perspective" can capture the perspective of God. He writes that "...there is no way in which a certain kind of anthropocentrism can be avoided."[57] This "certain kind of anthropocentrism" is Gustafson's way of characterizing the distinction between perspective and base. It is, in other words, simply an acknowledgment that a theological ethic is intended for human agents, not the divine agent, and therefore ought to be concerned with the perspective of human agents. But again, adopting a human perspective in this sense in no way justifies the centrality traditionally accorded humans. Gustafson recognizes with Heyd that humans invest the world with value, and he recognizes that the tradition has to a large extent supported this interpretation of humanity's place in creation, but that humans project themselves onto the world in this way is to some degree the very problem Gustafson is trying to address. Human agents are better seen, claims Gustafson, not as the "measure" of all things, but as the "measurers."[58]

In essence, then, Gustafson affirms with the tradition two of its three claims. He affirms that God is generally benevolent toward creation (as a whole), and that humans have a special moral agency relative to other species. However, he denies that God's purposes are principally directed toward human good; nor does he believe that God can be counted on to guarantee that God's purposes will coincide with human good, either in this life or the next. This distinguishing feature of his approach flows mainly from Gustafson's view of God.

For Gustafson,

> 'God' refers to the power that bears down upon us, sustains us, sets an ordering of relationships, provides conditions of possibilities for human activity and even a sense of direction.[59]

This view of God is based largely in his interpretation of religious experience, which he typically discusses in terms of "piety." The most significant implications of his understanding of piety for his view of God arise from its *dialectical character*. Gustafson says that piety is:

> a sense of awe evoked both by the radical contingency of all that is and by the reliable ordering of much of life...[60]

However, where other theological ethicists have emphasized the order or ordering of God's creation, Gustafson reminds us more strongly of its contingencies. By contingency, I believe Gustafson intends to convey both senses of the term outlined in chapter 1.[61] First, contingency refers to those chance events in natural and human history that have traditionally been treated as problems of theodicy. His emphasis on contingency in this sense serves to weaken traditional views of God's special concern for human well-being and flourishing. Second, he uses the term, as I do in this book, to lend more weight to human agency and decision-making than is sometimes done in the tradition. In other words, since God cannot be counted on to order creation in such a way as to guarantee human well-being and flourishing, human decision-making takes on even greater importance for determining future states of affairs. Both of these senses of contingency work to establish his impersonal view of God and to ground a more expansive moral domain or "whole."[62]

However, while God is viewed impersonally by Gustafson as a "power that bears down upon us," God continues to be viewed in agential terms, and thus is not considered utterly impersonal.[63] For example, Gustafson also discusses God as a power that "enables," "orders," creates "possibilities," and gives a "sense of direction" to human existence. This dialectical view of God also influences the language he uses. For example, God is typically discussed as "the Deity," and creation is characteristically discussed in more neutral terms as "nature." These terms also serve as a rhetorical device to constrain what he believes are natural, if self-serving, human tendencies to anthropomorphize God and reality generally.

In any case, the point is that God's purposes for the world, whatever they may be, do not in Gustafson's view remove or protect human persons from contingency, either in this life or ultimately in some form of salvation or afterlife. Thus, while there is obviously sufficient ordering in the natural world to permit humanity as a species to flourish, this ordering does not in Gustafson's estimation provide grounds for the belief that God will guarantee the well-being of the individual or, finally, of the entire species. Moreover, rather than trying to rationalize the contingency of human existence in some understanding of theodicy, it is embraced as part of the purposes of God to which humans must "consent" (though not passively) if they are finally not to destroy themselves and the conditions on which their lives depend. Again, from the perspective of theocentric ethics, the results of these qualifications are a de-centering of humans relative to the purposes of God for creation and a dramatically expanded moral domain. One commentator summarizes Gustafson's position this way:

> Since the human species seems doomed, and an afterlife is implausible, God's basic purpose must not be the well-being of the species or the individual. Retaining the assumption that God is beneficent as well as powerful, Gustafson reasons then that the purpose of God is the well-being of sentient life and in some sense the well-being of the nonsentient creation as well. Since we have a duty to conform to God's purposes...we should seek the good of all creation. Thus Gustafson establishes one basic part of his moral framework, a duty to serve the good of the whole.[64]

Discernment

Given these substantive views, Gustafson develops his "ethics from a theocentric perspective" by moving self-consciously from an "indicative" mode to the "imperative." He claims that:

> The theocentric perspective requires that the practical moral question be asked as follows: What is God enabling and requiring us to be and to do?[65]

In response, he says:

> The general answer...from a theocentric perspective is: we are to relate ourselves and all things in a manner (or in ways) appropriate to their relations to God.[66]

To give this formal answer substantive content for concrete actions or policies, that is, to decide which actions or policies under a given set of circumstances best relate us and all things in a manner appropriate to their relations to God, Gustafson develops a complex methodology that he characterizes as a process of "moral discernment."

Moral discernment, like McCormick's proportionalism, is basically teleological in structure and heavily weighted toward the prudential, especially when directed toward various public policy options. However, it may be fair to say that Gustafson gives even more weight to consequences than McCormick and, more importantly for addressing the problem of contingent future persons, that his methodology is more clearly agent-neutral. It is especially this latter qualification of his theocentric approach to value that permits a higher degree of impersonalism when making value judgments.

Discernment is primarily a way of making inferences from the experience of God in piety to an evaluation of human action and being. Agents make these inferences on the assumption that they can dramatically affect the conditions for life and well-being established and sustained by God, and that God intends humans to act in such a way as to promote and sustain these conditions.[67] Discernment begins with an evaluative description of the problem that includes relevant information, relevant causal factors, and an evaluation of the relevant agents' ability to affect outcomes. It seeks a wide perspective with respect to time and space, and tries to limit the instrumentalization of nature and of other living species. For evaluating public policies, input is invited from the wider public concerning the priority of values and principles to be used. Discernment is rational in a broad sense of the term and aimed at providing justification or reasons for one's decisions, choices, and actions. Finally, discernment asks what can be inferred about "the divine governance" of the world. Essentially, Gustafson says, we can discern that God enables us by providing the necessary conditions for life and that God requires us to sustain and develop these, but again, not simply for human well-being but for the well-being of all creation (the "whole").[68]

It is this last point, the "imperative" Gustafson believes is required by the theocentric perspective, that leads me to conclude his position is finally agent-neutral. This does not mean that discernment cannot and should not take account of the agent's perspective. "Sensitivities, affectivities, capacities for empathy, and imagination are important," Gustafson says.[69] Nevertheless, he is unremitting in his efforts to pull the agent beyond his or her own perspective, and even beyond that of the species as a whole. Said differently the agent's subjectivity must be acknowledged if it is to be overcome.

Now, it is consistent to hold a person-affecting approach to value that is both theocentric with respect to its base and agent-neutral with respect to the perspective agents ought to adopt when making value judgments.[70] Agent-neutrality, as Parfit defined the term, is impersonal in the sense that it requires all agents to adopt a common aim and thus an impartial perspective when making value judgments. It thus refers in this sense not to the way value is fundamentally connected to the world but to the perspective agents ought to adopt when justifying particular actions or policies. It is also consistent with an impersonal theocentric approach to value to hold that value is fundamentally person-affecting and yet ask agents to be concerned primarily with states of affairs rather than persons as such. This is because God is fundamentally the one "person" for whom the world is good in this approach to value; as human agents relate themselves and all things in a manner appropriate to their relations to God, it is a state of affairs at which they are aiming, irrespective of their own good or the good of those with whom they have special relations. Thus, should such a state of affairs ever be realized, it could only be judged impersonally from the perspective of the human agent.

Some Implications

I now want briefly to investigate how this impersonal theocentric approach to value might address the some of the cases we have examined thusfar. In expanding the moral domain to cover the "whole," it is clear that Gustafson is concerned to justify obligations to future people generally.[71] To the best of my knowledge, however, he does not address the problem of contingent future persons, so we must inquire whether it is possible to extend his thought in this direction. Key to this extension are the implications of his position for our notions of stewardship.

In contrast to the instrumentalization of creation under a personal theocentric approach to value, impersonal theocentrism holds that creation is intrinsically good; that is, creation is held to be good independent of its ability to advance or promote human ends. It is this view of creation that permits us to incorporate criteria into our theocentric ethics that are typically thought to be impersonal from the perspective of human agents. For example, if we are convinced that God's purposes include the maintenance of the necessary conditions for the sustenance of life in general, even if this maintenance comes at some cost to human persons living now or in the future, we could incorporate into our process of discernment the criteria of quantity and quality. These criteria would, I believe, permit us to make decisions concerning non-contingent and contingent future persons without regard to their status. What matters ethically is the quantity (number) of future persons and the quality of their lives as they are related to a God who evidently values creation as a whole, but seems to value particular individuals or species only insofar as they contribute to that whole.[72]

Thus, on the individual level, we could provide impersonal theocentric guidance to a teenager who is contemplating pregnancy, or to a couple who is considering their decision to implant a preembryo with a known genetic defect. We could argue that the teenager ought to wait because her child will have a better start in life, irrespective of who is born. And we could suggest that the couple refrain from deliberately bringing into existence a genetically affected child—for example, one who is deaf—because this child will not enjoy, all things considered, the quality of life a hearing child would. With an impersonal theocentric approach to value, we would make these judgments on the grounds that relating "ourselves and all things in a manner appropriate to their relations to God" implies that God is concerned to advance the quality and quantity of life overall, at least for the foreseeable future. On this view of God, the knowledge that life will eventually end does not give us warrants to be rash or careless with it now. All life is valued by God instrumentally for God's own ends, whatever they may be, but such life ought to be regarded as having intrinsic worth by human agents. Presumably, if the teenager did not wait or the couple implanted a genetically affected preembryo, the quality of life overall would not be maximized, and creation as a whole would be diminished to that extent. (On the other hand, this approach does not provide us with warrants to ignore those children who are born to teenage mothers or those that are born with genetic disorders of some sort; it merely provides us with warrants not to bring them into existence when such choices are in our powers.)

Similar conclusions can be offered for choices that are made on the collective level. Not only could an impersonal theocentric approach to value provide reasons to choose a conservation policy over a depletion policy, but unlike the personal theocentric approach it could provide reasons to choose one contingent future population over another, if we were convinced that the choice significantly affected the overall quality and quantity of life in the future. Again, assuming God's purposes include the well-being of creation in general, irrespective of who is born, we would need to decide which policy or action is most likely to promote this well-being.

On the collective level, however, I believe the impersonal theocentric approach to value would eventually suffer from the same problems that Parfit's strictly impersonal approach does. That is, it would be no better in resolving the theoretical problems of constraint that Parfit encountered at the limits of his thinking. The impersonal theocentric approach to value solves the problem of contingent future persons—Parfit's so-called non-identity problem—but it fails to tell us how to constrain quantity and quality when they conflict, as eventually they must on the collective level. However, a difference between Parfit's and Gustafson's positions can be noted in their reactions to this problem. Parfit is optimistic that the problem will eventually be solved, but Gustafson believes such choices are irredeemably conflicted and thus inevitably tragic.[73] By tragic, Gustafson means that the obliga tions arising from an impersonal theocentric approach to value may incur *morally justified* costs or harms that cannot be allocated equitably between generations or among individuals of a given generation. Moreover, as God cannot be counted on to bring ultimate justice for these people, they just suffer unfairly. Gustafson says:

> God has not ordered the world so that the pursuit of morally justified ends and means can be fulfilled without cost and sometimes suffering to particular groups of persons now and in the future.[74]

This tragic view of ethics will have implications for addressing the relation between the problem of contingent future persons and the problem of inter-generational justice resulting from the unequal allocation of costs and harms projected for HGP's three phase future. We consider these implications in the next and final chapter of this book.

NOTES

1. Richard A. McCormick, "Moral Theology Since Vatican II: Clarity or Chaos?," in *The Critical Calling: Reflections on Moral Dilemmas Since Vatican II*, Richard McCormick (Washington, DC: Georgetown University Press, 1989), p. 14.

2. James M. Gustafson, *Ethics from a Theocentric Perspective: Ethics and Theology*, vol. 2 (Chicago: The University of Chicago Press, 1984), p. 17.

3. To the best of my knowledge, the ethical implications of the problem of contingent future persons have not been analyzed either by theological ethicists or by moral theologians. I have found only passing references to it in the theological literature. See, for example, William P. George, "Regarding Future Neighbors: Thomas Aquinas and Concern for Posterity," unpublished paper, Loyola University, Chicago, n.d. and Ronald M. Green, "Future Generations, Obligations to," in *The Westminster Dictionary of Christian Ethics*, eds. James F. Childress and John Macquarrie (Philadelphia: The Westminster Press, 1986), pp. 242-243. Both of these sources contain one-sentence references to the problem. George's argument figures in the discussion below.

4. There are other authors I could examine, of course. Paul Ramsey, for example, appeals to a notion of covenant in the Judeo-Christian tradition to ground deontic protections for persons. He believes this notion can be universalized to include non-Christians, and perhaps even the larger natural world. See David H. Smith, "On Paul Ramsey: A Covenant-Centered Ethics for Medicine," in *Theological Voices in Medical Ethics*, eds. Allen Verhey and Stephen E. Lammers (Grand Rapids, MI: William E. Eerdmans Publishing Company, 1993), pp. 7-29. On the other hand, Stanley Hauerwas appeals to christological convictions as they are carried and historically conditioned in the Christian church to ground a more particularistic commitment to persons. See Stephen E. Lammers, "On Stanley Hauerwas: Theology, Medical Ethics, and the Church," in *Theological Voices in Medical Ethics*, eds. Allen Verhey and Stephen E. Lammers, pp. 57-77. Both of these positions, however, are fundamentally deontological, and with Heyd I am persuaded that such positions in principle cannot address the problem of contingent future persons. They can include actual or non-contingent future persons in the moral domain, but not potential or contingent future persons. See note 4, p. 123.

5. Recall that Heyd's "actual" persons can include future actual or non-contingent future persons.

6. Heyd, *Genethics: Moral Issues in the Creation of People*, p. 6.

7. Ibid., p. 87. See also p. 85 for an interesting discussion of the ways his position is not anthropocentric: because humans are the measure of all value, they cannot themselves be measured in value terms.

8. Though I have typically used the term "agent-centric" to describe his position, at the highest level of abstraction Heyd states that genethics is finally "evaluator-centric." Ibid., pp. 84-85.

9. Heyd says of his theologically-weighted introduction to genethics that, "This interepretation will be offered here as a metaphysical model for a secular ethical theory of procreation rather than as a an exegetical reading of the biblical text. The reader is again cautioned not to take this book as an essay in natural theology." Ibid., p. 3.

10. Ibid., p. 4.

11. Ibid., p. 83.

12. This is the distinction James Gustafson makes in discussing the theological basis for his particular approach to theocentric ethics. See his *Ethics from a Theocentric Perspective: Theology and Ethics*, vol. 1 (Chicago: The University of Chicago Press, 1981), p. 3.

13. This is not to say theologians have not tried to make such a claim. Consider, for example, a fascinating 17th century debate concerning God's alleged knowledge of "futuribles;" that is, knowledge not only of what is actually the case and of what in the future will actually be the case, but also of *would* be the case if certain choices were made by human agents. See John Mahoney, *The Making of Moral Theology: A Study of the Roman Catholic Tradition* (Oxford: Clarendon Press, 1989), p. 90.

14. McCormick, "Moral Theology Since Vatican II: Clarity or Chaos?," p. 14.

15. Quoted in Ibid., p. 14.

16. Mahoney, *The Making of Moral Theology*, p. 114.

17. McCormick, "Moral Theology Since Vatican II: Clarity or Chaos?," p. 14.

18. Ibid., pp. 14-15.

19. Richard A. McCormick, "Bioethics and Method: Where Do We Start?," p. 46.

20. This claim may be motivated in part by a desire to protect God from certain charges in the face of the problem of theodicy. See Gustafson, *Ethics from a Theocentric Perspective: Theology and Ethics*, vol. 1, p. 96.

21. Ibid., pp. 109-110. Quoted in Gordon D. Kaufman, "How is 'God' to be Understood in a Theocentric Ethics?," in *James M. Gustafson's Theocentric Ethics: Interpretations and Assessments*, eds. Harlan R. Beckley and Charles M. Swezey (Macon, GA: Mercer University Press, 1988), p. 18. Also, Gustafson credits Hans Jonas for the observation that Western ethics generally are anthropocentric.

Gustafson, *Ethics from a Theocentric Perspective: Theology and Ethics*, vol. 1, note 124, p. 81.

22. See Cahill, "On Richard McCormick: Reason and Faith in Post-Vatican II Catholic Ethics," p. 81.

23. If an action is absolutely prohibited on these grounds, the status of this prohibition is experientially-based and not principled, as McCormick's statement in the epigraph of this chapter implies.

24. Richard A. McCormick, "Pluralism in Moral Theology," in *The Critical Calling: Reflections on Moral Dilemmas Since Vatican II*, Richard A. McCormick (Washington, DC: Georgetown University Press, 1989), p. 134.

25. McCormick, "Moral Theology Since Vatican II: Clarity or Chaos?," p. 16.

26. Richard A. McCormick, "The Ethics of Reproductive Technology," in *The Critical Calling: Reflections on Moral Dilemmas Since Vatican II*, Richard A. McCormick (Washington, DC: Georgetown University Press, 1989), p. 330.

27. Richard A. McCormick, "Theology in the Public Forum," in *The Critical Calling: Reflections on Moral Dilemmas Since Vatican II*, Richard A. McCormick (Washington, DC: Georgetown University Press, 1989), pp. 200-201.

28. Richard A. McCormick, "The Ethics of Reproductive Technology," p. 345.

29. For a spectrum of contemporary positions on the status of the preembryo, see Norman M. Ford, *When did I Begin?: Conception of the Human Individual in History, Philosophy, and Science* (New York: Cambridge University Press, 1988). See also John A. Robertson, "Ethical and Legal Issues in Preimplantation Genetic Screening," *Fertility and Sterility* 57, no. 1 (January, 1992): 1-11; and John A. Robertson, "Legal and Ethical Issues Arising with Preimplantation Human Embryos," *Archives of Pathology and Laboratory Medicine* 116 (April, 1992): 430-435.

30. Kevin D. O'Rourke and Philip Boyle, eds., *Medical Ethics: Sources of Catholic Teachings*, 2nd ed. (Washington, DC: Georgetown University Press, 1993), pp. 106-109.

31. Ibid., pp. 106-109.

32. Richard A. McCormick, "The Ethics of Reproductive Technology," pp. 333-340.

33. Ibid., p. 346.

34. Roman Catholic bishops may continue to prohibit the production of *in vitro* preembryos on the grounds outlined above and in this way think to avoid the problem of contingent future persons; however, as we saw with our review of Parfit's and Heyd's arguments in the previous chapter, the problem can arise simply as a result of influencing the timing of conception. This is an option that is morally available to Roman Catholics, and thus the problem can challenge the

tradition's personalism in ways that do not require the allegedly illicit production of preembryos.

35. Richard A. McCormick, "Genetic Technology and Our Common Future," in *The Critical Calling: Reflections on Moral Dilemmas Since Vatican II*, Richard A. McCormick (Washington, DC: Georgetown University Press, 1989), p. 266.

36. For example, he justifies therapeutic research on minor subjects that does not directly benefit them. This position occasioned a well-known, extended debate between McCormick and Paul Ramsey. This debate is documented in Cahill, "On Richard McCormick: Reason and Faith in Post-Vatican II Catholic Ethics," note 12, pp. 101 and pp. 102-105.

37. McCormick, "Genetic Technology and Our Common Future," p. 267.

38. Ibid., p. 270.

39. See United States, Congress, Office of Technology Assessment, *Biomedical Ethics in U.S. Public Policy—Background Paper*, OTA-BP-BBS-105.

40. McCormick, "Genetic Technology and Our Common Future," p. 270.

41. McCormick, "The Ethics of Reproductive Technology," p. 344.

42. Ibid., p. 345.

43. See George, "Regarding Future Neighbors: Thomas Aquinas and Concern for Posterity," p. 21.

44. Ibid. Emphasis added.

45. See Mary Midgley's comments on anthropocentrism: "It [anthropocentrism] reduces all values other than those it favors...to instrumental status." Mary Midgley, "The Paradox of Humanism," in *James M. Gustafson's Theocentric Ethics: Interpretations and Assessments*, eds. Harlan R. Beckley and Charles M. Swezey (Macon, GA: Mercer University Press, 1988), p. 188.

46. See Genesis 1:26-31.

47. I say "might" because we cannot hold human agents responsible for all future states of creation.

48. Interestingly, God's presumed impartiality relative to future (actual or non-contingent) persons introduces an element of impersonalism into a person-affecting approach to value, though only on the collective level.

49. See Heyd, *Genethics: Moral Issues in the Creation of People*, n. 16, p. 239.

50. Gustafson, *Ethics from a Theocentric Prespective: Ethics and Theology*, vol. 2, p. 143.

51. Ibid.

52. Ibid., pp. 105.

53. Gustafson, *Ethics from a Theocentric Perspective: Theology and Ethics*, vol. 1, p. 18.

54. Ibid., p. 101.

55. Ibid., pp. 108.

56. Ibid.

57. Ibid., p. 115.

58. Ibid., p. 15.

59. Gustafson, *Ethics from a Theocentric Perspective: Theology and Ethics*, vol. 1, p. 264.

60. James M. Gustafson, "Afterword," in *James M. Gustafson's Theocentric Ethics: Interpretations and Assessments*, eds. Harlan R. Beckley and Charles M. Swezey (Macon, GA: Mercer University Press, 1988), p. 241. Emphasis added.

61. See above, p. 10.

62. Interestingly, where others might interpret this contingency as human "fatedness" and argue for a significantly reduced understanding of human agency, Gustafson's insistance that we acknowledge the contingency of the world—even to the point of including it in our understanding of God—permits him to develop expanded notions of human agency.

63. Gordon Kaufman has criticized Gustafson for holding an impersonal view of God, while at the same time he holds that God exercises some form of continuing agency with respect to creation. He thus seems to hold both an agential and a non-agential view of God at the same time. See Gordon D Kaufman, "How is God to be understood in a Theocentric Ethics?," in *James M. Gustafson's Theocentric Ethics: Interpretations and Assessments*, eds. Harlan R. Beckley and Charles M. Swezey (Macon, GA: Mercer University Press, 1988), pp. 13-38. I believe Kaufman is essentially correct on this point, though he may err too far to the one side of Gustafson's dialectical view of God when he claims that Gustafson presents a "totally impersonal" God that is not a "moral reality." Nevertheless, this tension in Gustafson's view of God cannot be denied. In personal conservation with Professor Gustafson, he stated that he is willing to live with the tension until "a better view comes along" that is more persuasive to him.

64. John P. Reeder, Jr., "The Dependence of Ethics," in *James M. Gustafson's Theocentric Ethics: Interpretations and Assessments*, eds. Harlan R. Beckley and Charles M. Swezey (Macon, GA: Mercer University Press, 1988), p. 120.

65. Gustafson, *Ethics from a Theocentric Perspective: Ethics and Theology*, vol. 2, p. 1.

66. Ibid., p. 2.

67. James M. Gustafson, *The Contributions of Theology to Medical Ethics*, The 1975 Pere Marquette Theology Lecture (Milwaukee, WI: Marquette University Press, 1975), pp. 13-15 and pp. 21-25.

68. Ibid., pp. 327-328.

69. Gustafson, *Ethics from a Theocentric Perspective: Theology and Ethics*, vol. 1, p. 327.

70. An agent-neutral approach to value judgments is also available to philosophical approaches to value, but not to Heyd's genethics. If genethics were agent-neutral, Heyd would in principle be unable to determine who is potential

and who is actual; such a determination is not required of an impersonal theocentric approach to value.

71. See, for example, Gustafson's critique of utilitarianism in *Ethics from a Theocentric Perspective: Ethics and Theology*, vol. 2, pp. 103-104 and his discussion of population problems, vol. 2, pp. 235-250.

72. Thus, while creation is instrinsically good for human agents, it presumably is good in an instrumental sense for God.

73. This view of tragedy arises formally with his pluralistic theory of good. With no overarching good to determine the ranking of subordinate goods, there will inevitably be conflicts that cannot be resolved.

74. James M. Gustafson, *Ethics from a Theocentric Perspective: Ethics and Theology*, vol. 2, p. 250.

6

The Challenge of Contingent Future Persons

> [A]gents are constrained by the goods they hold.[1]
>
> Michael Walzer

> [L]egislators...have a duty to consider consequences of their legislation that they can foresee, and not merely those that they desire.[2]
>
> Richard M. Hare

This book attempts to address a two-pronged problem relative to human genome research and future generations. It investigates how the US Human Genome Project is likely to affect future generations, and it asks what implications these effects hold for *evaluating* HGP and other far-reaching research efforts like it, particularly from a theological perspective. Based on a review in chapters 2 and 3 of HGP's basic and applied goals, its congressional and scientific justification, and a list of problems expected to arise with its applications, I suggested that HGP will likely affect future generations in two basic ways.

First, as a result of the congressional decision to maximize efficiency with HGP's "dedicated" or "big science" shape, and as a result of the "sequelae" associated with the research and development of HGP's spin-off applications, phase two of HGP's likely three-phase future could be considerably more costly for future generations than its other phases. Nancy Wexler's statement before Congress, quoted in the epigraph of chapter 3, summarizes this concern succinctly and forcefully.

> If we end in this halfway zone [that is, in phase two], then I think we are in deep trouble...[I]t could be a potentially disastrous situation, to sit in hiatus between detection and cure.[3]

Second, and equally troubling, because of the capabilities made possible by certain applications of HGP such as the preimplantation diagnosis of genetic disease and germ-line gene therapy, the existence, numbers, and identities of some future persons will be contingent on HGP's research. Chapters 4 and 5 reviewed2 the vexing problems such capabilities have created for moral philosophers and theological ethicists. Gregory Kavka sums up the issue this way:

The trouble with future people...is not that they do not exist yet, it is that they might not exist at all.

In order to limit the vulnerability of those contingent future persons who are actually born in the future, I argued that we ought to include contingent future persons in our moral deliberations directly, and that this requires either an impersonal approach to value such as that developed by Parfit or an impersonal theocentric approach to value such as that developed by Gustafson. In this chapter, then, I want to outline some of the implications of these two approaches to value for addressing the non-contingent future generations affected by HGP's allocation pattern, summarize the implications uncovered throughout the book for policy makers, and end by exploring some theologically-oriented implications that may merit further attention in the future. These implications are also part of what constitutes the "challenge" of contingent future persons.

EVALUATING HGP'S ALLOCATION PATTERN

I claimed above that HGP's two basic effects on future generations give rise to two corresponding problems. First, HGP's probable allocation pattern confronts us with a problem of inter-generational distributive justice, and second, the problem of contingent future persons holds that persons born as a result of HGP's applications can be neither harmed nor benefited by being brought into existence. I also suggested, following Congress' lead, that our general approach to evaluating these two problems ought to be consequentialist, albeit an appropriately qualified consequentialist approach.

One of the most significant qualifications of this general approach involves my claim that we need to adopt a very broad understanding of the relevant moral domain. I argued that we need to expand our moral domain in order to consider the public policy issues underlying the project's founding and shape, and to account for the considerable expanse of time that may be involved in researching and developing HGP's future biomedical applications. I also believe that we should adopt the qualifications Parfit developed in his treatment of non-contingent future generations; these qualifications limit the burdens related to the prediction and control of outcomes at the same time they yield a rather risk averse approach to the future. Thus, I believe we ought to use a collective form of consequentialism that is also pluralistic with respect to its theory of good, time-neutral with respect to future generations, and agent-neutral with respect to the perspective agents ought to adopt when making particular value judgments.

Having discussed these qualifications, however, we interrupted our evaluation of HGP's future allocation pattern until we addressed the problem of contingent future persons. This problem is logically prior to the question of inter-generational justice raised by the allocation pattern, for we are not trying to decide the rightness of a particular course of action but to develop a theory of value that tells us how to determine a right course of action. Here, I suggested, we must consider how to expand the moral domain in yet a third way. The problem of contingent future persons is perhaps unusual in that it requires us to consider our theory of value on the most fundamental level of which we can conceive. On this level, we must decide how we believe value is ultimately related or connected to the world and thus to ethics; moreover, there are only two options available to us, and they represent mutually exclusive and exhaustive approaches to value. We are forced to this choice in the effort to bring contingent future persons consistently into our moral considerations in light of the fact that it makes no sense to speak of harming or benefiting them by bringing them into existence.

In an effort to meet this challenge, we engaged in what David Heyd calls a "theoretical cost-benefit analysis." That is, having decided that the problem forces us to choose between two mutually exclusive and exhaustive approaches to value, we tried to decide which approach best addresses the problem in terms of consistency and plausibility, while at the same time implying many of our long-held common-sense moral intuitions. Unfortunately, we saw that this is not a simple choice, due in large part to the fact that the sheer novelty of the problem means that our intuitions relative to the moral status of contingent future persons are simply not settled. Nevertheless, as we progressed through the debate between Parfit and Heyd in chapter 4, and weighed the various options available to us, I argued that Parfit's impersonal approach to value is more compelling than Heyd's person-affecting approach. I was particularly persuaded by the ability of the impersonal approach, on both the individual and the collective levels, to include contingent future persons in the moral domain and to consider them directly as morally relevant variables in choices that involve the question of their existence or non-existence. This impersonal approach, I believe, renders contingent future persons less vulnerable to agents' interests than does a person-affecting approach that, like Heyd's, is anthropocentric and predominantly agent-relative. With Parfit, I am persuaded that anthropocentric approaches to value and to ethics are finally, and perhaps paradoxically, worse for people overall.

This conclusion, however, proved problematic for me as a theological ethicist, as I do not believe we can consistently adopt an impersonal approach to value in the sense put forward by Parfit and continue to hold a theistic view of God. This theological concern moved the discussion to a consideration of how the Christian tradition understands the fundamental nature of value, and thus of how it might address the problem of contingent future persons. I suggested that the tradition fundamentally conceives of value in "theocentric" terms, meaning that the existence of value is dependent on, and only on, the existence of God. This is formally a person-affecting position, though it is complicated by a further claim that the tradition makes concerning the central role given to human persons in the purposes of God. It is this latter claim that led me to characterize the tradition's approach to value as "personal theocentrism." Moreover, when used to address the problem of contingent future persons, it is this latter claim that renders it subject to the same problems I uncovered with Heyd's strictly anthropocentric approach.

Using McCormick's and George's work, I suggested ways the traditional approach to value can be extended to incorporate very wide understandings of the moral domain, exploring how it can be conceived in non-individualistic terms and showing how it can include creation generally. Nevertheless, while personal theocentrism provides the *motivation* to act morally toward future states of creation and the future persons who will live in these future states, I argue that it fails to provide concrete guidance to individual agents for choices involving contingent future persons and that it provides only very limited guidance to collective agents. Essentially, this approach to value gives agents no reason to believe that God favors one future person over another, which a personal theocentric position must do if agents are to be guided by it in particular choices that involve contingent future persons.[4]

Gustafson's view of God gives agents no more reason than the traditional view to believe God values one contingent future person over another; in fact, just the opposite. It gives us no reason to believe that God values any one future person or group of persons, contingent or non-contingent, over another. But this is its strength. Contingent and non-contingent future persons are valued by human agents, from an agent-neutral perspective, for their contribution to various states of affairs that are to be related and thus judged relative to God. The impersonal qualifications of God developed by Gustafson, coupled with his methodological claims both that agents ought to aim toward states of affairs and that they ought to adopt an agent-neutral perspective when making value judgments, permit us to incorporate many of Parfit's impersonal qualifications into what remains fundamentally a person-affecting approach to value. I distinguish his position from the traditional approach to value by calling it "impersonal theocentrism." Again, it is formally person-affecting in the sense that God's existence is held to be a necessary condition for the existence of value, but it is impersonal in that human agents, when making particular judgments ought to adopt an agent-neutral perspective and ought to be concerned fundamentally with states of affairs as these states of affairs are related to God.

Now, while my outline of this impersonal theocentric approach to value in chapter 5 can by no means be seen as an adequate justification of it (which would require an extended defense of Gustafson's views of God, theology, and methodology independent of their ability to address the perplexity of the problem of contingent future persons), I believe the characterization of the position at least demonstrates that such an approach can *directly* address the problem of contingent future persons. That is, it can permit us to *include* contingent future persons in the moral domain and consider them as a variable when we are trying to decide how we ought to relate them to God as part of a future state of affairs. Also, I believe it can include on less abstract levels of consideration the criteria of quality and quantity of life, though in contrast to Parfit these are valued as they are related to God. It thus "cashes out" ethically in ways that are very similar to Parfit's impersonal approach. Given these lower-level criteria, I now want to consider the implications of a secular commitment to an impersonal approach to value and a theological commitment to an impersonal theocentric approach to value for evaluating HGP's effects on *non-contingent* future generations.

Recall that HGP's congressional justification as a federal research policy depended principally on the claim that its "dedicated shape" is the most efficient means to reach certain basic and applied scientific ends. We noted also that Congress may have been concerned to avoid certain market-driven evils related to undirected genome research. In light of these considerations, I suggested that the "threshold" and "catalyst" metaphors might best describe HGP's justification as policy. In this interpretation, the unique or distinctive aspect of HGP is found not in its basic science as such but in its coordinated administrative structure and its public commitment to reach its goals within certain time and budget constraints.

The particular questions urged by this book concern HGP's "target population," and the implications this population may hold for HGP's overall justification. Who exactly is intended to benefit from this research? Contingent future persons, born as a result of the project's research cannot meaningfully be said to benefit from it, at least insofar as this research is used to bring them into existence. But it is meaningful to speak of non-contingent future persons being beneficiaries of HGP's research. Thus, due to the sheer scale of this massive undertaking, and due to the fact that HGP's biomedical applications are projected to enter routine medical use over the next one hundred to two hundred fifty years (with perhaps as much as one thousand years needed for researchers fully to interpret HGP's fifteen years of mapping and sequencing), thousands and perhaps millions of future people will find their lives affected by this research. But not all of these effects can be considered benefits.

As we have seen, these future applications are expected to enter the medical mainstream in a certain order or sequelae, with the diagnostic applications expected to come "on-line" in phase two, long before the majority of therapeutic applications expected in phase three. This is expected to aggravate considerably the therapeutic gap in clinical genetics. Again, I qualified this prediction by observing that we are discussing a cumulative or statistical product; thus, particular breakthroughs at any point in time are not ruled out as a possibility. It is also, of course, possible that funding will be interrupted and much of these predictions will be invalidated altogether. Nevertheless, ELSI researchers and other informed observers predict that funding will continue and that our society will thus be forced to bear massive social costs as a result of our commitment to pursue human genome research in a way that emphasizes efficiency.

It is anticipated that these costs will be especially high during phase two of HGP's projected three-phase future, and that they will originate primarily in a system-wide information overload problem which in turn stems from the large numbers of diagnostic tools being developed and entering routine medical use. ELSI researchers hope to constrain or minimize the problems resulting from this information by anticipating them with certain educational and policy efforts, but no one believes they can be completely avoided, and we saw that their constraint may have negative implications for HGP's overall efficiency.[5] Thus, because of the expected sequelae of HGP's unfolding applications, future generations will probably not benefit equally. Phase two generations may bear an inordinate share of HGP's social costs, while phase three and following generations may enjoy an inordinate share of its benefits. Again, this is a question of inter-generational distributive justice (a question which by itself does not address *intra*-generational access to these benefits, relative to rich and poor persons and nations). Thus, here we ask what, in general terms, a *prior* commitment to an impersonal approach or to an impersonal theocentric approach to value implies for evaluating this probable allocative pattern.

As noted above, an impersonal approach to value "is not about what would be good or bad for those people whom our acts affect."[6] This claim also holds for an impersonal theocentric approach to value, though on different grounds. Both Parfit's impersonal approach and Gustafson's impersonal theocentric approach are concerned with better or worse states of affairs. Thus, both of these positions could provide warrants to support the *unequal* allocation pattern projected for HGP. Parfit would need to judge whether human genome research contributes enough to the overall quality and quantity of life in the world to justify requiring future generations to bear the social costs of phase two. Gustafson, insofar as his position is adequately captured by the impersonal theocentric approach, could make the same determination relative to his interpretation of God's purposes for the world (which, like Parfit's approach, encompass much more than human health and well-being). Thus, both approaches could provide reasons why we and phase two generations ought to bear the costs and harms of genome research and development so that future generations may be benefited in phase three and following (again, these phases are statistical products, so a number of specific individuals in all phases will presumably benefit while others will bear certain costs related to this research). Moreover, to the extent that allocative or distributive fairness in this case is judged in terms of equal distribution of benefits and burdens, both approaches could provide reasons that seem to suggest that we and, especially, phase two generations will simply bear these costs unfairly.

This said, it is also important to recall that we are working on very high levels of abstraction when considering these various approaches to value. Thus, while both approaches could provide reasons to justify actions or policies that might have tragic or costly consequences for certain future individuals or groups of individuals, both could and would also justify lower level considerations in making the determination that a particular action or policy is justified. Thus, for example, Parfit's commitment to a time-neutral consequentialism suggests that discounting the interests of future generations, simply because they are future, could never be justified; the future persons living in phase two of HGP's future thus count no more and no less than currently living persons or persons who will exist in phase three and following. Thus, he might rule out HGP research altogether, all things considered, on the grounds that there are much more efficient ways to advance human health, judged impersonally, and at much less cost to current and to future generations. Likewise, while we saw at the end of the previous chapter that Gustafson acknowledges how his theocentric approach can justify tragic outcomes for specific individuals or even whole generations of individuals, he also argues that this tragedy provides no grounds for being morally cavalier. Moreover, both Parfit and Gustafson develop pluralistic theories of value. This pluralism implies that it is possible to constrain the maximization of HGP's efficiency by the very kinds of considerations being put forth by ELSI's researchers and other informed observers. It must be remembered, however, that these constraints will cut against the overall efficiency of the project.

Nevertheless, if both authors judged HGP worth pursuing all things considered, they could justify the unequal allocative pattern suggested by the three-phase future projected for HGP. The same unequal allocative pattern would be much harder to justify in a person-affecting approach, which is surely one of its appeals. However, lest we too easily dismiss the impersonal approaches of Parfit and Gustafson because of this implication, recall that Heyd incorporated a social discount rate in mapping his own person-affecting approach. A social discount rate would permit current agents to discount the costs incurred by future generations on the basis of a belief that their distance from us in time is morally relevant; thus, for example, Heyd could argue that intervening generations might develop technologies that would offset the bad effects of our choices. In any case, I believe the justification of a discount rate is easier for those approaches that adopt an agent-relative perspective when justifying particular judgments. In my estimation, however, an agent-relative perspective too easily permits self-interest to enter our moral deliberations. I follow Parfit's lead on this; that is, I believe a discount rate is not morally justified, and I advocate an agent-neutral perspective to help limit the role of self- or even generational-interest. These commitments, on the whole, serve to render the impersonal approach of Parfit and the impersonal theocentric approach of Gustafson more risk averse than is sometimes the case with methodologies that are weighted so heavily in consequentialist directions.

IMPLICATIONS FOR POLICY MAKERS

In summary, then, the implications of this study for evaluating HGP's three-phase future are at least these. First, a realization that contingent future persons are involved in the decision to fund and implement a multi-generational project such as HGP requires that policy makers confront the perplexing problem of contingent future persons. This confrontation should determine their fundamental approach to evaluating the probable outcomes of their choice. This fundamental approach determines whether and how contingent future persons are included in or excluded from the moral domain, and thus whether they are treated directly or indirectly. However, whatever approach is adopted, we must also wrestle with the methodological qualifications that are among the more common fare of moral philosophers and theological ethicists. Said differently, while our choice of an approach to value is extremely important, it does not do all the work for us ethically.

Thus, second, policy makers need to decide which of the qualifications we discussed above (and perhaps more that were not discussed) they will adopt and utilize. These qualifications are, as we saw, complex and numerous, and they essentially concern whether and how to constrain the maximization of the basic goods or values at stake in a given choice. So, once a fundamental approach to value is decided, policy makers should also consider whether their theories of value ought to be monistic or pluralistic; whether they will adopt an individual or collective form of consequentialism; whether they as agents ought to adopt an agent-relative or agent-neutral perspective when justifying particular choices; and, whether they should adopt a social discount rate or a time-neutral approach.

Third, they need to take a stance on the question of empirical uncertainty. The greater the uncertainty, the more difficult the evaluation will be. It may be decided that some of this empirical uncertainty cannot in principle be overcome, but whether this leads policy makers to be more or less risky must be determined on other grounds. Uncertainty, as Parfit's argument indicates, can be just as strong a reason to develop a risk averse position as a risk friendly position. If one adopts a risk averse position, but decides to move ahead in the face of uncertainty, one can institute periodic reviews of the policy's progress and consequences, recognizing that these reviews will raise the indirect costs of any policy and lower its overall efficiency.

IMPLICATIONS FOR THEOLOGICAL ETHICISTS

Lastly, I want to end by listing what I take are some implications of the study for theological ethicists. Following Heyd, I suggested above that a theistic understanding of God implies a person-affecting approach to value. That is, I do not believe that one can adopt an impersonal approach to value in the sense advocated by Parfit and consistently view God as the source or ground of value. I believe, however, that the impersonal qualifications developed for the theocentric approach to value are consistent given Gustafson's views of God and his methodology. And, though I have not tried to demonstrate this, I believe that the two theological responses I outlined above capture the only two approaches to value available to theists. I do not believe, for example, that a covenantal approach, which I take to be another way to conceptualize a personal theocentric approach to value, would offer any additional resources for responding to the problem. In fact, insofar as this approach is treated by ethicists in deontological terms, I believe it would be unable in principle to address the problem of contingent future persons. And insofar as it is treated in teleological terms, I believe it would enjoy the same strengths and suffer from the same limitations that McCormick's and George's natural law positions do. But I leave the defense of this claim for the future.

Beyond these methodological concerns, the theological treatment of the problem of contingent future persons suggests a number of other questions that are both theologically and ethically interesting, though these too I merely list for future work. Some of these concerns will be relevant for non-Christian theistic traditions as well.

Insofar as we are persuaded that the problem of contingent future persons is a genuine challenge for theological ethics,[7] relative to those contingent future persons who are actually born, the problem would seem to undermine the common theological affirmation that "life is a gift of God" for which, as the common correlate goes, they should be grateful. Certainly, it makes rational sense for any person to feel and to express gratitude to God for one's *continuing* life, that is, for the conditions that sustain it. In this case, there is a person who can be benefited by such gifts. It might even make sense, when suffering horribly, to be grateful for death. But, if one now exists as a person who was once a contingent future person, it may make no sense to thank God—or anyone else, for that matter—for bringing one into existence. If coming into existence cannot be a benefit to such a person, then existence cannot be a "gift" to the person who is brought into existence. This is not to suggest that the coming into existence of a contingent future person under the novel conditions described above is not good for God—indeed, this is the very basis of a theocentric evaluation of such actions—but it is to suggest that it may not be possible to view such a birth as a gift from the perspective of the person who is born.

If this claim is true, such a weakening of the bonds of gratitude between persons and God, and by implication between some parents and their children, may have adverse effects on the natural bonds that bind members of families together and that bind society more generally. On learning that my life is the product of a deliberate choice, rather than feeling gratitude toward my parents for having intentionally brought me into existence, I may rationally feel less "connected" to them. I might still owe them gratitude for my upbringing, but I would not owe them gratitude for life itself. On the other hand, if I have a miserable life due to the same choice my parents made in bringing me into existence, I also could not rationally blame them, for if they had chosen otherwise I would not exist. Analogously, the problem of contingent future persons may cast theodicy problems in a new light. Could I rationally complain to God if my life were miserable when any other choice would have meant that I would not exist?

The problem of contingent future persons also raises interesting questions for certain tradition-specific moral claims within the Christian tradition. For instance, official Roman Catholic teaching prohibits IVF technologies on a number of grounds, some of which were detailed above when I discussed McCormick's position. However, one of the reasons that IVF is prohibited might be undermined by the problem of contingent future persons. The recently approved "Religious and Ethics Directives for Catholic Health Services" states in Directive 40 that the use of donor gametes in IVF procedures (a so-called "heterologous fertilization") is "prohibited because it is contrary to the covenant of marriage, the unity of the spouses and the dignity proper to parents and the child."[8] It is, of course, the reference to the child that is problematic. If I am correct in my reading of this directive, any future child in such a case must be considered contingent, and thus its "dignity" cannot be offended by coming into existence using gametes donated by parents who are not married.

At stake in this sort of appeal is an important issue for the Roman Catholic Church. It holds that its official teaching must not contradict the natural law as interpreted by the Magisterium. Thus, while official teaching is not inconsistent when it refers to the harm preembryos or embryos might bear, since the Church holds that personhood begins at conception, it is problematic to refer to the harm a *contingent* future child might bear by coming into existence using IVF procedures. The bishops can consistently claim that bringing children into the world using donated gametes undermines the "covenant of marriage" (though this claim may not be true empirically), but I do not believe they can consistently appeal to the alleged "dignity" of contingent future children as a reason to prohibit IVF procedures. To be consistent, I believe they must either appeal to an impersonal theocentric conception of value, which would be in conflict with a personalistic interpretation of the natural law, or they must give up their reference to contingent future persons.

A theological concern can also be raised relative to the so-called "sources" used to develop the distinction between the personal and impersonal theocentric approaches to value. In developing these two positions, I relied heavily on the same sources that Gustafson does, and on the basis of these sources I argued that God values persons, present or future, impartially. However, many theologians who are more closely aligned with the tradition, and who could thus be classed in the personal theocentric category, appeal to various interpretations of "revelation" to suggest that God is partial toward certain groups. For example, theologians in the past have privileged Christians over Jews or Jews over Gentiles, and some contemporary theologians privilege various populations whom they judge to be oppressed, say on the basis of race, sex, class, or age. Now, while I do not believe these understandings of God's alleged partiality are warranted, they may be offered as grounds to undermine the problem of contingent future persons, for they seem to suggest reasons that could be produced to argue that God does in fact favor one future person over another.

For example, suppose we are facing a choice between two contingent future persons, A and B, whose future existence is dependent on our choice. Suppose also that we cannot choose both A and B. This is the kind of question faced, on an individual level, by Parfit's teenager or by Heyd's couple deciding when to conceive. Now suppose that revelation is used to support the claim that we must show partiality and select A over B. Such a revelation would place A in the category of non-contingent future persons ("actual" in Heyd's sense of the term), and thus the relevant agents would be obligated to consider A's interests, rights, or welfare under traditional, person-affecting constraints. It thus could be plausibly argued that the revelation could make a difference concerning which future person is to be considered non-contingent or contingent, actual or potential (that is, God could have chosen B over A), but *it would not change the moral status of contingent future persons as such*. I believe this observation holds whether the revelation concerns only particular future persons, or special groups of persons who are "chosen" in a covenant ceremony by God. For agents who accept the revelation as morally binding, it renders contingent or potential persons non-contingent or actual by designating whom God intends for the agent to bring into existence, but it would not render potential persons as such morally considerable in personal theocentric terms.

Before we leave this question, consider one other way that an appeal to some interpretation of revelation might undermine the problem of contingent future persons. Suppose it was argued that all contingent future persons who will exist in the future are in fact non-contingent to an eternal and omniscient God.[9] If I understand this argument properly, it does not propose that the category of contingent future persons be eliminated, but that it simply be considered empty or null from God's perspective. However, even if this argument could be made plausible, it does not help human agents. Knowing that whatever choice we make relative to contingent future persons is already foreseen by God gives us no basis whatsoever for making our choice. Such an argument would be equivalent to attempting an ethic from a perspective that only God can enjoy, and such an ethic, I submit, is not an ethic for human agents.

The last two questions focus on the impersonal theocentric approach as such. We can ask if anything would be gained by opting for an *utterly* impersonal (and thus non-theistic) view of God. This is a question worth speculating on, for it would remove some of the tensions we encountered with Gustafson's dialectical view of God and it might move us toward a religiously-based approach to value that is strictly impersonal in the sense advocated by Parfit. Assuming we would be motivated to construct such an argument, on the basis of an utterly impersonal view of God we could probably establish a cosmocentric or biocentric ethic. These substantive understandings of an impersonal approach to value might be more or less plausible than Gustafson's views, but I suspect they would cash out ethically in ways similar to Parfit's impersonal approach and to what I am calling an impersonal theocentric approach. That is, they would succeed in accounting for contingent future persons, but fail at the same theoretical limits Parfit's approach failed in trying to constrain conflicting evaluative criteria.

Finally, it must be admitted that while aspects of impersonal theocentrism are congruent with the Christian tradition generally, in large part the impersonal theocentric approach to value that I outlined above undermines the tradition as a whole. It preserves the important claim that God constitutes value, but it nevertheless is in great tension with the tradition's view that God's purposes for the world involve humans in some central way. Both of these views cannot be correct, and thus here too a choice is required. Assuming that I am correct in holding that an impersonal theocentric approach to value can address the problem of contingent future persons better than the personal theocentrism of the tradition, this alone is not adequate grounds to adopt an impersonal theocentric approach. On the other hand, the problem stands as a challenge to the tradition, and I am hoping that those theological ethicists and moral theologians who are working in it will take up its challenge. It should make for an interesting discussion.

NOTES

1. Michael Walzer, *Spheres of Justice: A Defense of Pluralism and Equality* (New York: Basic Books, Inc., Publishers, 1983), p. 7.

2. R. M. Hare, "Public Policy in a Pluralist Society," in *Embryo Experimentation*, eds. Peter Singer, Helga Kuhse, Stephen Buckle, Karen Dawson, and Pascal Kasimba (Cambridge: Cambridge University Press, 1990), p. 186.

3. See above, p. 68.

4. A personal theocentric approach holds that God values persons and non-contingent future persons. Contingent future persons are valued only when they become persons in the fullest sense of the term. Below, I ask how a belief that God is in fact partial with regard to certain future persons or populations might affect this analysis.

5. That these questions were not settled before the project actually began is a legitimate concern. We can at least be thankful that the ELSI program was conceived and executed in order to investigate and to plan for the adverse affects of HGP that have been and will be identified. We can also be grateful that a periodic review process—designed to assess particular developments incrementally or as they develop—was built into HGP. In any case, a periodic review may be the best that can be done with such uncertainties. Congress could have banned the research outright until such time as all the long-range cost and risk factors were identified and evaluated, but such a ban might itself be difficult to justify without the experience that could only be gained from going forward with the research. Moreover, such a ban could likely be enforced only in

government funded research facilities. Given the financial incentives of HGP's expected applications, private laboratories could and probably would have moved ahead with the research.

6. Parfit, *Reasons and Persons*, p. 386.

7. There have been a number of attempts, none successful in my judgment, in the philosophical literature to undermine the problem on the level of definition. See, for example, Parfit's Appendix G, "Whether Causing Someone to Exist can Benefit this Person," Parfit, *Reasons and Persons*, pp. 487-490.

8. See Directive No. 40, National Conference of Catholic Bishops, "Religious and Ethical Directives for Catholic Health Services," *Origin* 24, no. 27 (December 15, 1994): 457.

9. This objection was raised in conversation with the author by Professor H. Tristram Engelhardt, Jr.

SELECTED BIBLIOGRAPHY

ACOG Committee Opinion: Committee on Ethics, Number 136, April 1994. "Preembryo Research: History, Scientific Background, and Ethical Considerations. *International Journal of Gynecology and Obstetrics* 45 (1994): 291-301.

Agius, Emmanuel. *The Rights of Future Generations: In Search of an Intergenerational Ethical Theory*. Ph.D. diss., Catholic University of Leuven, 1986.

Anderson, W. French. "Genetics and Human Malleability." *Hastings Center Report* 20 (January/February 1990): 21-24.

________. "Genetic Therapy." In *The New Genetics and the Future of Man*, ed. Michael Hamilton, 109-32. Grand Rapids, MI: Eerdmans Publishing Company, 1972.

________. "Human Gene Therapy: Scientific and Ethical Considerations." *Journal of Medicine and Philosophy* 10 (1985): 275-91.

________. "Human Gene Therapy: Why Draw a Line." *Journal of Medicine and Philosophy* 14 (1989): 681-93.

________. "Prospects for Human Gene Therapy." *Science* 226 (1984): 401-09.

________. "Reflections: Of Hope and of Concern." Editorial. *Human Gene Therapy* 2 (1991): 193-94.

________. "Uses and Abuses of Human Gene Transfer." Editorial. *Human Gene Therapy* 3 (1992): 1-2.

Anderson, W. French, and John C. Fletcher. "Gene Therapy in Human Beings: When is It Ethical to Begin?" *New England Journal of Medicine* 303 (1980): 1293-97.

Andolsen, Barbara Hilkert, Christine E. Gudorf, and Mary D. Pellauer, eds. *Women's Consciousness, Women's Conscience: A Reader in Feminist Ethics*. San Francisco: Harper & Row, Publishers, 1985.

Annas, George J. "Mapping the Human Genome and the Meaning of Monster Mythology." The Randolph W. Thrower Symposium on Genetics and the Law. Emory University, Atlanta, GA, 1990.

Annas, George J., and Sherman Elias, eds. *Gene Mapping: Using Law and Ethics as Guides*. New York: Oxford University Press, 1992.

________, eds. "The Major Social Policy Issues Raised by the Human Genome Project." In *Gene Mapping: Using Law and Ethics as Guides*, ed. George J. Annas and Sherman Elias, 3-17. New York: Oxford University Press, 1992.

Anscombe, G. E. M. "Modern Moral Philosohpy." *Philosophy* 33 (January 1958): 1-19.

Antonarakis, Stylianos E. "Diagnosis of Genetic Dis orders at the DNA Level." *The New England Journal of Medicine* 320 (19 January 1989): 153-63.

Aranson, Peter H. *The Political Economy of Science and Technology Policy.* Unpublished Manuscript. Emory University, 1988, Atlanta.

Armour, Leslie. "The Origin of Values." In *Ethics and Justification*, edited by Douglas Odegard, 177-93. Edmonton, Alberta, Canada: Academic Printing and Publishing, 1988.

Audi, Robert. "Theology, Science, and Ethics in Gustaf son's Theocentric Vision." In James M. Gustafson's *Theocentric Ethics: Interpretations and Assessments*, ed. Harlan R. Beckley and Charles M. Swezey, 159-83. Macon, GA: Mercer University Press, 1988.

Bales, R. Eugene. "Act-Utilitarianism: Account of Right-making Characteristics or Decision-making Procedure." *American Philosophical Quarterly* 8 (July 1971): 257-65.

Barrell, Bart. "DNA Sequencing: Present Limitations and Prospects for the Future." *The FASEB Journal* 5 (January 1991): 40-45.

Barry, Brian. "Circumstances of Justice and Future Gen erations." In *Obligations to Future Generations*, ed. R. I. Sikora and Brian Barry, 204-48. Philadelphia: Temple University Press, 1978.

________. "Rawls on Average and Total Utility: A Comment." *Philosophical Studies* 31 (1977): 317-25.

Beauchamp, Tom L. "The Moral Adequacy of Cost-benefit Analysis as the Basis for Government Regulation of Research." In *Ethical Issues in Government, Philosophical Monographs Third Annual Series*, ed. Norman E. Bowie, 163-75. Philadelphia: Temple University Press, 1981.

Beauchamp, Tom L., and James F. Childress. *Principles of Biomedical Ethics.* 3d ed. New York: Oxford University Press, 1989.

Beckley, Harlan R., and Charles M. Swesey, eds. *James M. Gustafson's Theocentric Ethics: Interpretations and Assessments.* Macon, GA: Mercer University Press, 1988.

Bellah, Robert N., Richard Madsen, William M. Sullivan, Ann Swidler, and Steven M. Tipton. *The Good Society.* New York: Alfred A. Knopf, Inc., 1991.

Berg, Paul. "All Our Collective Ingenuity Will Be Needed." *The FASEB Journal* 5 (January 1991): 75.

Berger, Edward M., and Bernard M. Gert. "Genetic Dis orders and the Ethical Status of Germ-line Therapy." *The Journal of Medicine and Philosophy* 16 (1991): 667-83.

Bickman, Stephen. "Future Generations and Contemporary Ethical Theory." *Journal of Value Inquiry* 15, no. 2 (1981): 169-77.

SELECTED BIBLIOGRAPHY

Billings, Paul R., Cassandra L. Smith, and Charles R. Cantor. "New Techniques for Phyical Mapping of the Human Genome." *The FASEB Journal* 5 (January 1991): 28-34.

Bishop, Jerry E. "Unnatural Selection." *National Forum* (Spring 1993): 27-29.

Bodmer, Walter F. "Gene Clusters, Genome Organization, and Complex Phenotypes. When the Sequence is Known, What Will It Mean?" The William Allen Memorial Award Address. *American Journal of Human Genetics* 33 (1981): 664-82.

Bond, E. J. "The Justification of Moral Judgments." In *Ethics and Justification*, ed. Douglas Odegard, 55-63. Edmonton, Alberta, Canada: Academic Printing and Publishing, 1988.

Bonnicksen, Andrea. "Genetic Diagnosis of Human Embryos." Special Supplement. *Hastings Center Report* 22 (July-August 1992): S5-11.

Botstein, David. "Construction of a Genetic Linkage Map in Man Using Restriction Fragment Length Polymor phisms." *American Journal of Human Genetics* 32 (1980): 314-31.

Buckle, Stephen. "Arguing from Potential." In *Embryo Experimentation*, ed. Peter Singer, Helga Kuhse, Stephen Buckle, Karen Dawson, and Pascal Kasimba, 90-108. Cambridge: Cambridge University Press, 1990.

Byrne, John. "A Critique of Beauchamp and Braybrooke-Schotch." In *Ethical Issues in Government, Philosophical Monographs Third Annual Series*, ed. Norman E. Bowie, 198-217. Philadelphia: Temple University Press, 1981.

Cahill, Lisa Sowle. "On Richard McCormick: Reason and Faith in Post-Vatican II Catholic Ethics." In *Theological Voices in Medical Ethics*, ed. Allen Verhey and Stephen E. Lammers, 78-105. Grand Rapids, MI: William E. Eerdmans Publishing Company, 1993.

Callahan, Daniel. "Ethical Responsibility in Science in the Face of Uncertain Consequences." *Annals of the New York Academy of Sciences* 265 (1976): 1-11.

________. "Religion and the Secularization of Bioethics." Special Supplement. *Hastings Center Report* (July/August 1990): 2-4.

________. "What Obligations Do We Have to Future Generations?" In *Responsibilities to Future Generations: Environmental Ethics*, ed. Ernest Partridge, 73-85. Buffalo, NY: Prometheus Books, 1981.

Cantor, Charles R. "Orchestrating the Human Genome Project." *Science* 248 (6 April 1990): 49-51.

Caplan, Arthur. "Mapping Morality: Ethics and the Human Genome Project." In *If I Were a Rich Man Could I Buy a Pancreas?*, Arthur Caplan, 118-42. Bloomington: Indiana University Press, 1992.

Caplan, Arthur L. "If Gene Therapy is the Cure, What is the Disease?" In *Gene Mapping: Using Law and Ethics as Guides*, ed. George J. Annas and Sherman Elias, 128-41. New York: Oxford University Press, 1992.

Caskey, C. Thomas, and Victor A. McKusick. "Medical Genetics." *JAMA* 263 (May 16 1990): 2654-56.

Catholic Health Association of the United States. *Human Genetics: Ethical Issues in Genetic Testing, Counseling, and Therapy*. St. Louis, MO: Catholic Health Association of the United States, 1990.

Center for Biologics Evaluation and Research, and Food and Drug Administration. "Points to Consider in Human Somatic Cell Therapy and Gene Therapy." *Human Gene Therapy* 2 (1991): 251-56.

Cook-Deegan, Robert M. "Social and Ethical Implications of Advances in Human Genetics." *Southern Medical Journal* 83 (August 1990): 879-82.

________. "The Alta Summit, December 1984." *Genomics* 5 (1989): 661-63.

________. "The Human Genome Project: The Formation of Federal Policies in the United States, 1986-1990." In *Biomedical Politics*, ed. K. E. Hanna, 99-168. Washington, D.C.: National Academy Press, 1991.

Critser, Elizabeth S. "Preimplantation Genetics: An Overview." *Archives of Pathology and Laboratory Medicine* 116 (April 1992): 383-87.

Cutter, Mary Ann G., et al. *Mapping and Sequencing the Human Genome: Science, Ethics, and Public Policy*. Colorado Springs, CO: BSCS and American Medical Association, 1992.

Dahl, Robert A. *Democracy and Its Critics*. New Haven: Yale University Press, 1989.

Daniels, Norman. "Introduction." In *Reading Rawls: Critical Studies on Rawls' "A Theory of Justice"*, ed. Norman Daniels, xxxi-liv. Stanford Series in Philosophy. Stanford, CA: Stanford University Press, 1989.

________. "Preface." In *Reading Rawls: Critical Studies on Rawls' "A Theory of Justice"*, ed. Norman Daniels, xiii-xxx. Stanford Series in Philosophy. Stanford, CA: Stanford University Press, 1989.

________, ed. *Reading Rawls: Critical Studies on Rawls' "A Theory of Justice"*. Stanford Series in Philosophy. Stanford, CA: Stanford University Press, 1989.

Dawson, Karen. "Introduction: An Outline of Scientific Aspects of Embryo Research." In *Embryo Experimentation*, ed. Peter Singer, Helga Kuhse, Stephen Buckle, Karen Dawson, and Pascal Kasimba, 3-13. Cambridge: Cambridge University Press, 1990.

________. "Introduction." In *Embryo Experimentation*, ed. Peter Singer, Helga Kuhse, Stephen Buckle, Karen Dawson, and Pascal Kasimba, xiii-xvi. Cambridge: Cambridge University Press, 1990.

De George, Richard T. "The Environment, Rights, and Future Generations." In *Ethics and the Problems of the 21st Century*, ed. K. E. Goodpastor and K. M. Sayre, 93-105. Notre Dame: University of Notre Dame Press, 1979.

Draper, Elaine. "Genetic Secrets: Social Issues of Medical Screening in a Genetic Age." Special Supplement. *Hastings Center Report* 22 (July-August 1992): S15-18.

Dulbecco, Renato. "A Turning Point in Cancer Research: Sequencing the Human Genome." *Science* 231 (7 March 1986): 1055-56.

Dustira, Alicia K. "The Funding of Basic and Clinical Biomedical Research." In *Biomedical Research: Collaboration and Conflict of Interest*, ed. Roger J. Porter and Thomas E. Malone, 33-56. Baltimore, MD: The Johns Hopkins University Press, 1992.

Elias, Sherman, and George J. Annas. "Somatic and Germ- line Gene Therapy." In *Gene Mapping: Using Law and Ethics as Guides*, ed. George J. Annas and Sherman Elias, 142-54. New York: Oxford University Press, 1992.

Engelhardt, H. Tristram, Jr. *Bioethics and Secular Humanism: The Search for a Common Morality*. Philadelphia: Trinity Press International, 1991.

________. *The Foundation of Bioethics*. New York: Oxford University Press, 1986.

Epstein, Suzanne L. "Regulatory Concerns in Human Gene Therapy." *Human Gene Therapy* 2 (1991): 243-49.

Farley, Edward. "Theocentric Ethics as a Genetic Argument." In James M. Gustafson's *Theocentric Ethics: Interpretations and Assessments*, ed. Harlan R. Beckley and Charles M. Swezey, 39-58. Macon, GA: Mercer University Press, 1988.

Farley, Margaret A. "Feminist Theology and Bioethics." In *Women's Consciousness, Women's Conscience: A Reader in Feminist Ethics*, ed. Barbara Hilkert Andolsen, Christine E. Gudorf, and Mary D. Pellauer, 285-305. San Francisco: Harper & Row, Publishers, 1985.

Feinberg, Joel. "The Rights of Animals and Unborn Generations." In *Responsibilities to Future Generations: Environmental Ethics*, ed. Ernest Partridge, 139-50. Buffalo, NY: Prometheus Books, 1981.

Ferguson-Smith, M. A. "European Approach to the Human Gene Project." *The FASEB Journal* 5 (January 1991): 61-65.

Flack, Harley E., and Edmund D. Pellegrino, eds. *African-American Perspectives on Biomedical Ethics*. Washington, D.C.: Georgetown University Press, 1992.

Fletcher, John C. "Ethical Issues In and Beyond Prospective Clinical Trials of Human Gene Therapy." *Journal of Medicine and Philosophy* 10 (1985): 293-309.

Ford, Norman M. *When Did I Begin?: Conception of the Human Individual in History, Philosophy and Science*. New York: Cambridge University Press, 1988.

Frankena, William K. *Ethics*. 2d ed. Prentice-Hall Foundations in Philosophy Series. Englewood Cliffs, NJ: Prentice-Hall, Inc., 1973.

Gaylin, Willard. "Fooling with Mother Nature." *Hastings Center Report* (January/February 1990): 17-21.

George, William P. "Regarding Future Neighbors: Thomas Aquinas and Concern for Posterity." Unpublished Paper. Loyola University, n.d., Chicago.

Glendon, Mary Ann. *Abortion and Divorce in Western Law: American Failures, European Challenges*. Cambridge, MA: Harvard University Press, 1987.

Glover, Jonathan. "Matters of Life and Death." *The New York Review of Books* (30 May 1985), 19-20.

Golding, Martin P. "Obligations to Future Generations." In *Responsibilities to Future Generations: Environmental Ethics*, ed. Ernest Partridge, 61-72. Buffalo, NY: Prometheus Books, 1981.

Goodin, Robert E. "Utility and the Good." In *A Campanion to Ethics*, ed. Peter Singer, 241-48. Cambridge, MA: Basil Blackwell, 1991.

Goodpastor, K. E., and K. M. Sayre, eds. *Ethics and the Problems of the 21st Century*. Notre Dame: University of Notre Dame Press, 1979.

Green, Eric D., and Robert H. Waterston. "The Human Genome Project: Prospects and Implications for Clinical Medicine." *Journal of the American Medical Association* 266 (9 October 1991): 1966-75.

Green, Ronald M. "Future Generations, Obligations To." In *The Westminster Dictionary of Christian Ethics*, ed. James F. Childress and John Macquarrie, 242-43. Philadelphia: The Westminster Press, 1986.

Grobstein, Clifford, and Michael Flower. "Gene Therapy: Proceed with Caution." *The Hastings Center Report* 14 (April 1984): 13-17.

Gunnemann, Jon P. "Michael Walzer and the Relativity of Justice." Unpublished Paper. Emory University, 1983, Atlanta.

Gustafson, James M. "Afterword." In James M. Gustafson's *Theocentric Ethics: Interpretations and Assessments*, ed. Harlan R. Beckley and Charles M. Swezey, 241-54. Macon, GA: Mercer University Press, 1988.

________. *Can Ethics Be Christian?* Chicago: The University of Chicago Press, 1975.

________. *The Contributions of Theology to Medical Ethics*. The 1975 Pere Marquette Theology Lecture. Milwaukee, WI: Marquette University Press, 1975.

________. *Ethics from a Theocentric Perspective: Ethics and Theology*. Vol. 2. Chicago: The University of Chicago Press, 1984.

________. *Ethics from a Theocentric Perspective: Theology and Ethics*. Vol. 1. Chicago: The University of Chicago Press, 1981.

________. "Response." In James M. Gustafson's *Theocentric Ethics: Interpretations and Assessments*, ed. Harlan R. Beckley and Charles M. Swezey, 203-24. Macon, GA: Mercer University Press, 1988.

Habermas, Jurgen. *The Structural Transformation of the Public Sphere: An Inquiry Into a Category of Bourgeois Society.* Translated by Thomas Burger, with assistance by Frederick Lawrence. Cambridge, MA: The MIT Press, 1989.

________. *The Theory of Communicative Action.* 2 Vol. Translated by Thomas McCarthy. Boston: Beacon Press, 1984-7.

Hare, R. M. "Public Policy in a Pluralist Society." In *Embryo Experimentation*, Eds Peter Singer, Helga Kuhse, Stephen Buckle, Karen Dawson, and Pascal Kasimba, 183-94. Cambridge: Cambridge University Press, 1990.

________. "Rawls' Theory of Justice." In *Reading Rawls: Critical Studies on Rawls' "A Theory of Justice"*, ed. Norman Daniels, 81-107. Stanford Series in Philosophy. Stanford, CA: Stanford University Press, 1989.

Heller, Jan C. "The U.S. Human Genome Project: Mapping the Moral Boundaries of an Interim Ethic." Unpublished Paper. Emory University, 1992, Atlanta.

Heyd, David. *Genethics: Moral Issues in the Creation of People.* Berkeley: University of California Press, 1992.

Holbrook, Daniel. *Qualitative Utilitarianism.* Lanham, MD: University Press of America, 1988.

Holtzman, Neal A. *Proceed with Caution.* Baltimore, MD: Johns Hopkins University Press, 1989.

Holtzman, Neil A. "Recombinant DNA Technology, Genetic Tests, and Public Policy." *American Journal of Human Genetics* 42 (1988): 624-32.

Hoose, Bernard. "Gene Therapy: Where to Draw the Line." *Human Gene Therapy* 1 (1990): 299-306.

Hopper, David H. *Technology, Theology, and the Idea of Progress.* Louisville, KY: Westminster/John Knox Press, 1991.

Ikawa, Yoji. "Human Genome Efforts in Japan." *The FASEB Journal* 5 (January 1991): 66-69.

Johnson, Conrad D. *Moral Legislation: A Legal-political Model for Indirect Consequentialist Reasoning.* Cambridge Studies in Philosophy. Cambridge: Cambridge University Press, 1991.

Jonas, Hans. *The Imperative of Responsibility: In Search of an Ethics for the Technological Age.* Translated by Hans Jonas, in collaboration with David Herr. Chicago: The University of Chicago Press, 1984.

________. "Technology and Responsibility: The Ethics of an Endangered Future." In *Responsibilities to Future Generations: Environmental Ethics*, ed. Ernest Partridge, 23-36. Buffalo, NY: Prometheus Books, 1981.

Juengst, Eric T. "Germ-line Gene Therapy: Back to Basics." *The Journal of Medicine and Philosophy* 16 (1991): 587-92.

________. "The Human Genome Project and Bioethics." *Kennedy Institute of Ethics Journal* 1 (March 1991): 71-74.

_______. "Human Genome Research and the Public Interest: Progress Notes from an American Science Policy Experiment." *American Journal of Human Genetics* 54 (1994): 121-28.

Kaufman, Gordon D. "How is `God' to Be Understood in a Theocentric Ethics?" In James M. Gustafson's *Theocentric Ethics: Interpretations and Assessments*, ed. Harlan R. Beckley and Charles M. Swezey, 13-35. Macon, GA: Mercer University Press, 1988.

Kavka, Gregory. "The Futurity Problem." In *Obligations to Future Generations*, ed. R. I. Sikora and Brian Barry. 180-203. Philadelphia: Temple University Press, 1978.

Kavka, Gregory S. "The Paradox of Future Individuals." *Philosophy and Public Affairs* 11 (1982): 93-112.

Kimura, Rihito. "Religious Aspects of Human Genetic Information." In *Human Genetic Information: Science, Law, and Ethics*, eds. Derek Chadwick, Greg Bock, and Julie Welan, 148-55. Chichester, UK: John Wiley & Sons Ltd, 1990.

Kolata, Gina. "Unlocking the Secrets of the Genome." *The New York Times* 30 November 1993, B5-6.

Kulse, Helga, and Peter Singer. "Individuals, Humans, and Persons: The Issue of Moral Status." In *Embryo Experimentation*, ed. Peter Singer, Helga Kuhse, Stephen Buckle, Karen Dawson, and Pascal Kasimba, 65-75. Cambridge: Cambridge University Press, 1990.

Lammers, Stephen E. "On Stanley Hauerwas: Theology, Medical Ethics, and the Church." In *Theological Voices in Medical Ethics*, ed. Allen Verhey and Stephen E. Lammers, 57-77. Grand Rapids, MI: William E. Eerdmans Publishing Company, 1993.

Lammers, Stephen E., and Allen Verhey, eds. *On Moral Medicine: Theological Perspectives in Medical Ethics*. Grand Rapids, MI: William B. Eerdmans Publishing Company, 1987.

Lappé, Marc. "Ethical Issues in Manipulating the Human Germ Line." *The Journal of Medicine and Philosophy* 16 (1991): 621-39.

_______. "The Limits of Genetic Inquiry." *Hastings Center Report* 17 (August 1987): 5-10.

_______. "Long-term Implications of Mapping & Sequencing the Human Genome: Ethical and Philosophical Implications." In *Mapping Our Genes: Federal Genome Projects: How Vast? How Fast?* Contractor Reports, Vol. 1., United States, Congress, Office of Technology Assessment, 233-303. Springfield, VA: National Technical Information Service, 1988.

Lebacqz, Karen. "Feminism and Bioethics: An Overview." *Second Opinion* 17 (October 1991): 11-25.

Lee, Thomas F. *The Human Genome Project: Cracking the Genetic Code of Life*. New York: Plenum Press, 1991.

Levit, Katharine R., Helen C. Lazenby, Cathy A. Cowan, and Suzanne W. Letsch. "National Health Expenditures, 1990." *Health Care Financing Review* 13 (Fall 1991): 29-54.

Lindblom, Charles E. Politics and Markets: *The Worlds Political-economic Systems*. New York: Basic Books, Inc., Publishers, 1977.

Lyons, David. *Forms and Limits of Utilitarianism*. Oxford: The Clarendon Press, 1965.

McCarthy, Cormac. *All the Pretty Horses*. New York: Vintage International, Vintage Books, 1992.

McCormick, Richard A. "Bioethics and Method: Where Do We Start?" In *On Moral Medicine: Theological Perspectives in Medical Ethics*, ed. Stephen E. Lammers and Allen Verhey, 45-54. Grand Rapids, MI: William B. Eerdmans Publishing Company, 1987.

________. *The Critical Calling: Reflections on Moral Dilemmas Since Vatican II*. Washington, D.C.: Georgetown University Press, 1989.

________. "Genetic Technology and Our Common Future." In *The Critical Calling: Reflections on Moral Dilemmas Since Vatican II*, Richard A. McCormick, 261-72. Washington, D.C.: Georgetown University Press, 1989.

________. "Moral Theology Since Vatican II: Clarity or Chaos?" In *The Critical Calling: Reflections on Moral Dilemmas Since Vatican II*, Richard A. McCormick, 3-24. Washington, D.C.: Georgetown University Press, 1989.

________. "Pluralism in Moral Theology." In *The Critical Calling: Reflections on Moral Dilemmas Since Vatican II*, Richard A. McCormick, 131-46. Washington, D.C.: Georgetown University Press, 1989.

________. "Theology in the Public Forum." In *The Critical Calling: Reflections on Moral Dilemmas Since Vatican II*, Richard A. McCormick, 191-208. Washington, D.C.: Georgetown University Press, 1989.

________. "Therapy or Tampering? The Ethics of Reproductive Technology and the Development of Doctrine." In *The Critical Calling: Reflections on Moral Dilemmas Since Vatican II*, Richard A. McCormick, 329-52. Washington, D.C.: Georgetown University Press, 1989.

MacIntyre, Alasdair. "Utilitarianism and Cost-benefit Analysis: An Essay on the Relevance of Moral Philosophy to Bureaucratic Theory." In *Values in the Electric Power Industry*, ed. Kenneth Sayre, 217-37. Notre Dame: University of Notre Dame Press, 1977.

McKusick, Victor A. "Current Trends in Mapping Human Genes." *The FASEB Journal* 5 (January 1991): 12-20.

________. "The Human Genome Organization: History, Purposes, Membership." *Genomics* 5 (1989): 385-87.

________. "The Human Genome Project: Plans, Status, and Applications in Biology and Medicine." In *Gene Mapping: Using Law and Ethics as Guides*, ed.

George J. Annas and Sherman Elias, 18-42. New York: Oxford University Press, 1992.

________. "Mapping and Sequencing the Human Genome." *New England Journal of Medicine* 320 (6 April 1989): 910-15.

McNaughton, David, and Piers Rawling. "Honoring and Promoting Values." *Ethics* 102 (July 1992): 835-43.

Mahoney, John. *The Making of Moral Theology: A Study of the Roman Catholic Tradition*. Oxford: Clarendon Press, 1989.

Matter, Beth. "Mapping the Human Genome: Will It Solve the Mystery of Life?" *Vanderbilt Magazine* (Fall 1993), 17-21.

Mauron, Alex, and Jean-Marie Th,voz. "Germ-line Engineering: A Few European Voices." *The Journal of Medicine and Philosophy* 16 (1991): 649-66.

Midgley, Mary. "The Paradox of Humanism." In James M. Gustafson's *Theocentric Ethics: Interpretations and Assessments*, ed. Harlan R. Beckley and Charles M. Swezey, 187-99. Macon, GA: Mercer University Press, 1988.

Miller, A. Dusty. "Human Gene Therapy Comes of Age." *Nature* 357 (11 June 1992): 455-60.

Miller, Peter. "On Justifying Morality." In *Ethics and Justification*, ed. Douglas Odegard, 197-212. Edmonton, Alberta, Canada: Academic Printing and Publishing, 1988.

Monk, Marilyn. "Embryo Research and Genetic Disease." *New Scientist* 6 January 1990, 56-59.

Moseley, Ray, Lee Crandall, Marvin Dewar, David Nye, and Harry Ostrer. "Ethical Implications of a Complete Human Gene Map for Insurance." *Business and Professional Ethics Journal* 10, no. 4 (1991): 69-82.

Munson, Ronald, and Lawrence Davis, H. "Germ-line Gene Therapy and the Medical Imperative." *Kennedy Institute of Ethics Journal* 2, no. 2 (1992): 137-58.

Murray, Thomas H. "Ethical Issues in Human Genome Research." *The FASEB Journal* 5 (January 1991): 55-60.

Narveson, Jan. "Utilitarianism and New Generations." *Mind* 76 (1967): 62-72.

National Conference of Catholic Bishops, "Religious and Ethical Directives for Catholic Health Services," *Origin* 24, no. 27 (December 15, 1994): 449-462.

National Research Council, Commission on Life Sciences, Board of Basic Biology, and Committee on Mapping and Sequencing the Human Genome. *Mapping and Sequencing the Human Genome*. Washington, D.C.: National Academy Press, 1988.

Nelkin, Dorothy. "Genetics and Social Policy." *Bulletin of the New York Academy of Medicine* 68 (January-February 1992): 135-43.

Nelkin, Dorothy, and Laurence Tancredi. *Dangerous Diagnostics: The Social Power of Biological Information*. New York: Basic Books, 1989.

Nolan, Kathleen. "First Fruits: Genetic Screening." Special Supplement. *Hastings Center Report* 22 (July-August 1992): S2-4.

O'Rourke, Kevin D., and Philip Boyle, eds. *Medical Ethics: Sources of Catholic Teachings*. 2d ed. Washington, D.C.: Georgetown University Press, 1993.

Odegard, Douglas. "Introduction." In *Ethics and Justification*, ed. Douglas Odegard, 1-15. Edmonton, Alberta, Canada: Academic Printing and Publishing, 1988.

________, ed. *Ethics and Justification*. Edmonton, Alberta, Canada: Academic Printing and Publishing, 1988.

Palmer, Julie Gage. "Liability Considerations Presented by Human Gene Therapy." *Human Gene Therapy* 2 (1991): 235-42.

Parfit, Derek. "Overpopulation and the Quality of Life." In *Applied Ethics*, ed. Peter Singer, 145-64. Oxford: Oxford University Press, 1986.

________. *Reasons and Persons*. New York: Oxford University Press, 1984.

Partridge, Ernest, ed. *Responsibilities to Future Generations: Environmental Ethics*. Buffalo, NY: Prometheus Books, 1981.

Passmore, John. *Man's Responsibility for Nature: Ecological Problems and Western Traditions*. New York: Charles Scribner's Sons, 1974.

Pearson, Mark L., and Dieter S"ll. "The Human Genome Project: A Paradigm for Information Management." *The FASEB Journal* 5 (January 1991): 35-39.

Pellegrino, Edmund D. "The Metamorphosis of Medical Ethics: A 30-year Retrospective." *JAMA* 269 (March 3 1993): 1158-62.

Pettit, Philip. "Consequentialism." In *A Campanion to Ethics*, 230-40. Cambridge, MA: Basil Blackwell, 1991.

Post, Stephen G. "Selective Abortion and Gene Therapy: Reflections on Human Limits." *Human Gene Therapy* 2 (1991): 229-33.

President's Commission for the Study of Ethical Problems in Medicine and Biomedical and Behavioral Research. *Splicing Life: A Report on the Social and Ethical Issues of Genetic Engineering with Human Beings*. Washington, D.C.: U.S. Government Printing Office, 1982.

Proctor, Robert N. "Genomics and Eugenics: How Fair is the Comparison?" In *Gene Mapping: Using Law and Ethics as Guides*, ed. George J. Annas and Sherman Elias, 57-93. New York: Oxford University Press, 1992.

Ramsey, Paul. "Genetic Therapy: A Theologian's Response." In *The New Genetics and the Future of Man*, ed. Michael Hamilton, 157-75. Grand Rapids, MI: Eerdmans Publishing Company, 1972.

Rawls, John. *A Theory of Justice*. Cambridge, MA: The Belknap Press of Harvard University Press, 1971.

Reeder, John P., Jr. "The Dependence of Ethics." In James M. Gustafson's *Theocentric Ethics: Interpretations and Assessments*, ed. Harlan R. Beckley and Charles M. Swezey, 119-37. Macon, GA: Mercer University Press, 1988.

Ripley, Randall B., and Grace A. Franklin. *Congress, the Bureaucracy, and Public Policy*. 5th ed. Pacific Grove, CA: Brooks/Cole Publishing Company, 1991.

Robbins, Robert J. "Challenges in the Human Genome Project: Progress Hinges on Resolving Database and Computational Factors." *IEEE Engineering in Medicine and Biology* (March 1992): 25-34.

Robertson, John A. "Ethical and Legal Issues in Preimplantation Genetic Screening." *Fertility and Sterility* 57 (January 1992): 1-11.

_______. "Genetic Alteration of Embryos: The Ethical Issues." In *Genetics and the Law* III, ed. Aubrey Milunsky and George J. Annas. 115-27. New York: Plenum Press, 1985.

_______. "Legal and Ethical Issues Arising with Preimplantation Human Embryos." *Archives of Pathology and Laboratory Medicine* 116 (April 1992): 430-35.

Rossiter, Belinda J. F., and C. Thomas Caskey. "Molecular Studies of Human Genetic Disease." *The FASEB Journal* 5 (January 1991): 21-27.

Rothman, Barbara Katz. "Not All the Glitters is Gold." Special Supplement. *Hastings Center Report* 22 (July-August 1992): S11-15.

Rothstein, Mark A. *Medical Screening and the Employee Health Cost Crisis*. Washington, D.C.: Bureau of National Affairs, 1989.

Rowen, Henry. "The Role of Cost-benefit Analysis in Policy Making." In *Cost Benefit Analysis and Water Polution Policy*, ed. Henry M. Peskin and Eugene P. Seskin. Washington, D.C.: Urban Institute, 1975.

Rowley, Peter T. "No Limits to Genetic Inquiry." Letter to the Editor. *Hastings Center Report* 18 (April/May 1988): 42-43.

Scheffler, Samuel. "Agent-Centered Restrictions, Rationality, and the Virtues." In *Consequentialism and Its Critics*, ed. Samuel Scheffler, 243-60. Oxford: Oxford University Press, 1988.

_______. "Introduction." In *Consequentialism and Its Critics*, ed. Samuel Scheffler, 1-13. Oxford: Oxford University Press, 1988.

_______. *The Rejection of Consequentialism: A Philosophical Investigation of the Considerations Underlying Rival Moral Theories*. Oxford: Clarendon Press, 1982.

Schell, Jonathan. *The Fate of the Earth*. New York: Alfred A. Knopf, 1982.

Schwartz, Thomas. "Obligations to Posterity." In *Obligations to Future Generations*, ed. R. I. Sikora and Brian Barry, 3-13. Philadelphia: Temple University Press, 1978.

Shuster, Evelyne. "Determinism and Reductionism: A Greater Threat Because of the Human Genome Project?" In *Gene Mapping: Using Law and Ethics as Guides*, ed. George J. Annas and Sherman Elias, 115-27. New York: Oxford University Press, 1992.

Sikora, R. I., and Brian Barry, eds. *Obligations to Future Generations*. Philadelphia: Temple University Press, 1978.

Singer, Peter, Helga Kuhse, Stephen Buckle, Karen Dawson, and Pascal Kasimba, eds. *Embryo Experimentation*. Cambridge: Cambridge University Press, 1990.

Singer, Peter, and Karen Dawson. "IVF Technology and the Argument from Potential." In *Embryo Experimentation*, ed. Peter Singer, Helga Kuhse, Stephen Buckle, Karen Dawson, and Pascal Kasimba, 76-89. Cambridge: Cambridge University Press, 1990.

Sinsheimer, Robert. "The Santa Cruz Workshop, May 1985." *Genomics* 5 (1989): 954-65.

Slote, Michael. *Beyond Optimizing: A Study of Rational Choice*. Cambridge, MA: Harvard University Press, 1989.

________. *Common-sense Morality and Consequentialism*. International Library of Philosophy. London: Routledge & Kegan Paul, 1985.

Smart, J. J. C. "An Outline of a System of Utilitarian Ethics." In *Utilitarianism: For and Against*, J. J. C. Smart and Bernard Williams, 3-74. Cambridge: Cambridge University Press, 1973.

Smith, David H. "On Paul Ramsey: A Covenant-centered Ethics for Medicine." In *Theological Voices in Medical Ethics*, ed. Allen Verhey and Stephen E. Lammers, 7-29. Grand Rapids, MI: William E. Eerdmans Publishing Company, 1993.

Sonnefeld, Sally T., Daniel R. Waldo, Jeffrey A. Lemieux, and David R. McKusick. "Projections of National Health Expenditures Through the Year 2000." *Health Care Financing Review* 13 (Fall 1991): 1-27.

Sowden, Lanning. "Parfit on Self-interest, Common-Sense Morality and Consequentialism: A Selective Critique of Parfit's `Reasons and Persons'" *The Philosophical Quarterly* 36, no. 145: 514-35.

Starr, Paul. *The Social Transformation of American Medicine: The Rise of a Sovereign Profession and the Making of a Vast Industry*. New York: Basic Books, Inc., Publishers, 1982.

Steinbock, Bonnie. *Life Before Birth: The Moral and Legal Status of Embryos and Fetuses*. New York: Oxford University Press, 1992.

Stephens, J. Claiborne, Mark L. Cavanaugh, Margaret I. Gradie, Matrin L. Mador, and Kenneth K. Kidd. "Mapping the Human Genome: Current Status. *Science* 237 (12 October 1990): 237-44.

Strickland, Stephen P. *Politics, Science, and Dread Disease: A Short History of United States Medical Research Policy*. A Commonwealth Fund Book. Cambridge, MA: Harvard University Press, 1972.

Suzuki, D., and P. Knudtson. *Genethics*. Cambridge, MA: Harvard University Press, 1989.

Taylor, Charles. "The Diversity of Goods." In *Utilitarianism and Beyond*, ed. Amartya Sen and Bernard Williams, 129-44. Cambridge: Cambridge University Press, 1982.

Tong, Rosemarie. *Ethics in Policy Analysis*. Prentice-Hall Series in Occupational Ethics. Englewood Cliffs, NJ: Prentice-Hall, Inc., 1986.

Toulmin, Stephen Edelston. *An Examination of the Place of Reason in Ethics*. Cambridge: Cambridge University Press, 1958.

U.S. Department of Energy. *Human Genome: 1989-90 Program Report*. DOE/ER-0446P. [Washington, D.C.]: U.S. Department of Energy, 1990.

United States, Congress, and Office of Technology Assessment. *Biomedical Ethics in U.S. Public Policy—Background Paper*. OTA-BP-BBS-105. Washington, D.C.: U.S. Government Printing Office, 1993.

United States, Congress, House, and Committee on Goverment Operations. *Designing Genetic Information Policy: The Need for an Independent Policy Review of the Ethical, Legal, and Social Implications of the Human Genome Project*. House Report 102-478. Washington, D.C.: U.S. Government Printing Office, 1992.

United States, Congress, House, Space Committee on Science, and Technology, and Subcommittee on International Scientific Cooperation. *Hearing on International Cooperation in Mapping the Human Genome*. 19 October 1989, 2325 Rayburn House Office Building. Washington, D.C.: Government Printing Office, 1989.

United States, Congress, Senate, Science and Transportation Committee on Commerce, and Technology and Space Subcommittee on Science. *Hearing on the Human Genome Initiative and the Future of Biotechnology*. S. Hrg. 101-528. Washington, D.C.: U.S. Government Printing Office, 1989.

United States, Congress, Senate, Committee on Energy and Natural Resources, and Subcommittee on Energy Research and Development. *Hearing on the Human Genome Project*. S. Hrg. 101-894. Washington, D.C.: U.S. Government Printing Office, 1990.

United States, Congress, House, Committee on Energy and Commerce, and Subcommittee on Oversight and Investigations. *Hearing on the OTA Report on the Human Genome Project*. Serial No. 100-123. Washington, D.C.: U.S. Government Printing Office, 1988.

United States, Congress, Senate, Committee on Energy and Natural Resources, and Subcommittee on Energy Research and Development. *Hearings on Department of Energy National Laboratory Cooperative Research Initiatives Act*. S. Hrg. 100-602, Pt. 1. Washington, D.C.: U.S. Government Printing Office, 1988.

United States, Congress, and Office of Technology Assessment. *Mapping Our Genes; Genome Projects: How Big, How Fast?* OTA-BA-373. Washington, D.C.: U.S. Government Printing Office, 1988.

United States, Congress, and House. Representative Michael A. Andrews of Texas. "Mapping the Human Genome: The Most Important Biological

Project in the History of Science." 101st Congress. *Congressional Record* 135 (27 April 1989): E 1418-19.

United States, Congress, and Office of Technology Assessment. *Medical Testing and Health Insurance*. OTA-H-384. Washington, D.C.: U.S. Government Printing Office, 1988.

_______. *Medical Testing and Health Insurance*. OTA-H-384 Summary. Washington, D.C.: U.S. Government Printing Office, 1988.

United States, Congress, House, and Representative Obey of Wisconsin. "Speaking to the Human Genome Project Funded by the Labor, Health, and Human Services and Education Appropriations Bill." 101st Congress. *Congressional Record* 136 (19 July 1990): H 5002-03.

United States, Congress, and Office of Technology Assessment. Technologies for *Detecting Heritable Mutations in Human Beings*. OTA-H-298 Summary. Washington, D.C.: U.S. Government Printing Office, 1986.

_______. *Technologies for Detecting Heritable Mutations in Human Beings*. OTA-H-298. Washington, D.C.: U.S. Government Printing Office, 1986.

United States, Congress, Senate, and Committee on Energy and Natural Resources. *Workshop on Human Gene Mapping*. 100-71. Washington, D.C.: U.S. Government Printing Office, 1988.

United States Department of Health and Human Services, Public Health Service, and National Institutes of Health. "Points to Consider in the Design and Submission of Human Somatic-cell Gene Therapy Protocols." Eve K. Nichols. In *Human Gene Therapy*, 195-208. Cambridge, MA: Harvard University Press, 1986.

United States Department of Health and Human Services, Public Health Service, National Institutes of Health, United States Department of Energy, Office of Energy Research, and Office of Health and Environmental Research. *Understanding Our Genetic Inheritance; the U.S. Human Genome Project: The First Five Years, FY 1991-1995*. DOE/ER-0452P. Washington, D.C.: National Institutes of Health, U.S. Department of Health and Human Services; and U.S. Department of Energy, 1990.

Vaccaro, Vincent. "Cost-benefit Analysis and Public Policy Formation." In *Ethical Issues in Government*, ed. Norman E. Bowie, 146-62. Philosophical Monographs Third Annual Series. Philadelphia: Temple University Press, 1981.

Vaughan, Christopher C., Bruce L. R. Smith, and Roger J. Porter. "The Contributions of Biomedical Science and Technology to U.S. Economic Competitiveness." In *Biomedical Research: Collaboration and Conflict of Interest*, ed. Roger J. Porter and Thomas E. Malone, 57-76. Baltimore, MD: The Johns Hopkins University Press, 1992.

Veatch, Robert M., ed. *Cross Cultural Perspectives in Medical Ethics: Readings*. Boston: Jones and Bartlett Publishers, 1989.

Verhey, Allen. "On James M. Gustafson: Can Medical Ethics Be Christian?" In *Theological Voices in Medical Ethics*, ed. Allen Verhey and Stephen E. Lammers, 30-56. Grand Rapids, MI: William E. Eerdmans Publishing Company, 1993.

Verhey, Allen, and Stephen E. Lammers, eds. *Theological Voices in Medical Ethics*. Grand Rapids, MI: William E. Eerdmans Publishing Company, 1993.

Walzer, Michael. *Spheres of Justice: A Defense of Pluralism and Equality*. New York: Basic Books, Inc., Publishers, 1983.

Warren, Mary. "Do Potential People Have Rights?" In *Obligations to Future Generations*, ed. R. I. Sikora and Brian Barry, 14-30. Philadelphia: Temple University Press, 1978.

Watson, James D. "The Human Genome Project: Past, Present, and Future." *Science* 248 (6 April 1990): 44-48.

Watson, James D., and Eric T. Juengst. "Doing Science in the Real World: The Role of Ethics, Law, and the Social Sciences in the Human Genome Project." In *Gene Mapping: Using Law and Ethics as Guides*, ed. George J. Annas and Sherman Elias, xv-xix. New York: Oxford University Press, 1992.

Watson, J[ames] D., and F[rancis] H. C. Crick. "Molecular Structure of Nucleic Acids: A Structure for Deoxyribose Nucleic Acid." *Nature* 171 (25 April 1953): 737-38.

Watson, James Dewey, and Robert Mullan Cook-Deegan. "Origins of the Human Genome Project." *The FASEB Journal* 5 (January 1991): 8-11.

Weatherall, D. J. *The New Genetics and Clinical Practice*. 3rd ed. Oxford: Oxford University Press, 1991.

Wertz, Dorothy C. "Ethical and Legal Implications of the New Genetics: Issues for Discussion." *Social Science and Medicine* 35, no. 4 (1992): 495-505.

White, Ray, and C. Thomas Caskey. "Genetic Predisposition and the Human Genome Project: Case Illustrations of Clinical Problems." In *Gene Mapping: Using Law and Ethics as Guides*, eds. George J. Annas and Sherman Elias, 173-85. New York: Oxford University Press, 1992.

Williams, Bernard. "A Critique of Utilitarianism." In *Utilitarianism: For and Against*, J. J. C. Smart and Bernard Williams, 77-150. Cambridge: Cambridge University Press, 1973.

________. *Ethics and the Limits of Philosophy*. Cambridge, MA: Harvard University Press, 1985.

Wuthnow, Robert. *Meaning and Moral Order: Explorations in Cultural Analysis*. Berkeley: University of California Press, 1987.

Yesley, Michael S., compiler, with the assistance of Michael R. J. Roth. *ELSI Bibliography: Ethical, Legal, and Social Implications of the Human Genome Project*. DOE/ER-0591. Washington, D.C.: U.S. Department of Energy, Office of Energy Research, 1993.

Young, Iris Marion. *Justice and the Politics of Difference*. Princeton, NJ: Princeton University Press, 1990.

Zimmerman, Burke K. "Human Germ-line Therapy: The Case for Its Development and Use." *The Journal of Medicine and Philosophy* 16 (1991): 593-612.